American Institute of Chemical Engineers

Sensors

Topical Conference at the
2007 AIChE Annual Meeting

November 4-9, 2007
Salt Lake City, Utah, USA

Printed from e-media with permission by:

Curran Associates, Inc.
57 Morehouse Lane
Red Hook, NY 12571
www.proceedings.com

ISBN: 978-1-60423-837-2

ISBN: 978-1-60423-837-2

American Institute of Chemical Engineers
Proceedings
Three Park Avenue
New York, NY 10016-5991
Phone: 212-591-8100

www.aiche.org

American Institute of Chemical Engineers

Sensors
2007

TABLE OF CONTENTS

426a Monolithic Tunnel Junctions For Single Molecule Sensor Devices

Brian G. Willis, University of Delaware, Colburn Laboratory, 150 Academy Street, Newark, DE 19716 and Rahul Gupta, Chemical Engineering, University of Delaware, Colburn Laboratory, 150 Academy Street, Newark, DE 19716.

Molecular electronics is a developing area of scientific research where long term objectives are to create active molecular devices that act as computing or sensing devices, for example. Over the past 10 years significant progress has been made in the measurement of single molecule electrical properties such as electrical conductance, electronic energy states, and even single molecule vibration spectroscopy using tunneling electrons. The later phenomena, inelastic electron tunneling spectroscopy (IETS), is potentially a very useful sensing method because it provides a vibrational fingerprint of molecules, has single molecule sensitivity, and has the advantage that it is a solid state technique with potential for large scale, low cost manufacturing using microfabrication methods. Arguably, the most successful experiments to date have been carried out using scanning tunneling microscopy methods, which are not practical outside of the laboratory. In this paper, we present a novel approach to IETS sensors using monolithically nanofabricated tunnel junctions grown using atomic layer deposition (ALD). ALD is used to grow the sensing nanoelectrodes and control the tunnel junction spacing from field emission (> 2 nm electrode separation) to direct tunneling (< 2 nm separation). ALD will be shown to be an ideal nanoelectrode growth technique due to the sub-nm growth precision, ultra-clean growth environment (of special importance for electrode materials other than gold), the potential for controlling electrode atomic structure, and the potential for scalability to manufacturing. In-situ electrical measurements during the tunnel junction growth will be presented to show the transition from field emission to direct tunneling, the effects of temperature variation and thermal expansion, and chemical effects due to oxidation of the electrodes. The electrodes will be shown to have several advantages over existing molecular measurement techniques including stability to voltages up to +/-5 V, and insensitivity to vibration. Moreover, the monolithic tunnel junction approach allows molecular trapping experiments to be carried out in real time. Implications for the extension of these devices to monolithic IETS sensors will be discussed.

Comparison of chemical etching variations for Tapered Optical Fibers

Yinni Yu, D. Keith Roper

Department of Chemical Engineering, 3290 MEB, University of Utah, Salt Lake City, Utah 84112

Introduction

Single mode Tapered Optical Fibers (TOF) are employed as tips in Near-field Scanning Optical Microscopy (NSOM) imaging and spectroscopy applications. NSOM provides optical resolution that surpasses the diffraction limit of ~200 nm in conventional microscopy, because the optical resolution for NSOM is a function of the tip diameter through which a particular electromagnetic wavelength is transmitted [1][2]. Scattering NSOM, where delocalized electrons are excited at the tip, can also be used with IR or Raman spectroscopy to obtain spectroscopic information about a specimen [1][3]. There are two common ways to fabricate TOFs, CO_2 laser-heated pulling [2][4] and HF-silicone oil mixture etching [2][5], Burgos et al. [2] reported that the former produced cone angles of 25°-30° and tip size of about 200 nm; whereas the latter had cone angle of 90° and tip size of about 150 nm. TOF tip diameter and cone angle are important parameters in NSOM applications that affect resolution and light transmission. At wavelength of 633nm, chemically etched fiber had 10-40 times higher light transmission than laser-pulled fiber of the same aperture size due to bigger cone angle from chemical etching [2]. The purpose of this paper has been to compare chemical etching methods used to produce TOFs, and investigate the systematic variations in concentration and time parameters that yield desired TOF tip size and cone angles reproducibly. We adapted a modified procedure from Maruyama et al. [3] to reproducibly fabricate TOFs with tip sizes of 70 - 80 nm and cone angles between 20° and 26°.

Experimental Section

We used single mode optical fiber (F-SA, Newport, Irvine, CA) that consisted of a silica (Si) core doped with GeO_2 (3.4 μm diameter), clad by fused-silica with diameter of 125±1 μm, and coated by an acrylate buffer to a final outer diameter of 245 ±12 μm. The acrylate buffer was mechanically stripped ~0.5 cm from the end of the fiber to expose the cladding using a mechanical stripper (T06S13, Thorlabs, Newton, NJ). Fiber etch using silicone/HF in a polypropylene container is shown in Schematic 1. Polypropylene does not react with the HF acid. The meniscus at the silicone oil/HF interphase yielded the tapered shape at the tip of the fibers as discussed further in discussion section. In order to study the effect of the different etching time intervals, the meniscus height (the length of fiber in contact with the HF) was maintained at 2 ± 0.5 mm. Four chemical etching variations are compared, the first step in all four methods was a silicone oil/hydrofluoric acid (HF) etch in which 49% stock HF and an overlayer of silicone oil (Dow Corning) were used [5]. All etching experiments were done at room temperature.

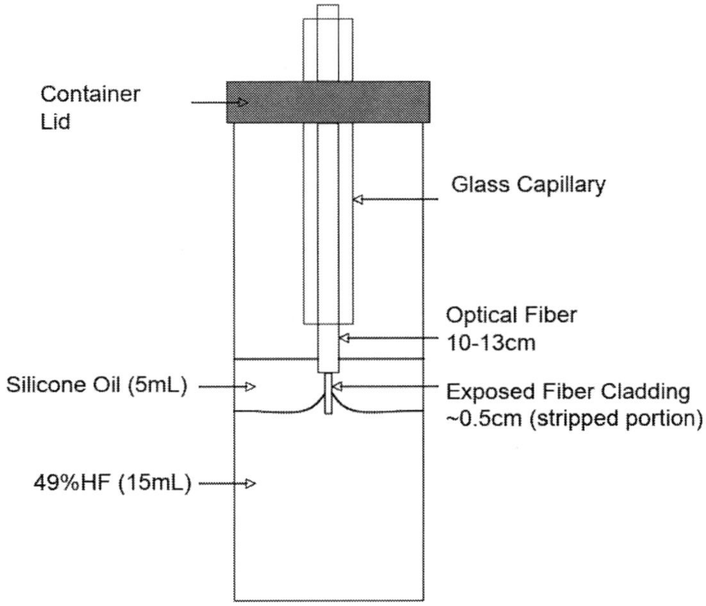

Schematic 1. Optical fiber etch in silicone oil/49%HF inside polypropylene tube, fiber was supported by threading it through a glass capillary (*not drawn to scale*).

Variation I: submerge the fiber with silicone oil/49% HF for successively longer periods (40 -120 minutes) in the setup as shown in Schematic 1. After the etch, the fibers were removed from the etchant and dipped in a running bath of deionized H_2O, then gently washed with acetone and methanol to wash off the organic silicone oil. Rinsing under a faucet of deionized H_2O was avoided because the oncoming pressure can break the tip of the TOF.

Variation II: this variation employed a second etch, using 49% HF solution diluted four-fold with deionized H_2O to 12.25% [6]. Fibers etched with Variation I were washed as mentioned above, then immediately immersed in the second etch for 40 and 60 minutes. Note that all second etch procedures do not require silicone oil, and the fiber should be washed with acetone and methanol to avoid introducing unwanted silicone oil into the solution. After the second etch is completed, the fibers were dipped again in running bath of deionized H_2O.

Variation III: the third variation was silicone oil/49% HF followed by second etch with buffered HF (BHF), which had the volumetric ratio of 4:1:3 (40 wt% NH_4F: 49% HF: H_2O), for 40 and 60-minute time points. The BHF ratio of 4:1:3 was adopted from the procedure for selective etching done by Maruyama et al. [4]. Similar to Variation II, the fibers were properly washed prior to the second etch, which was done without the silicone oil.

Variation IV: this was the triple-etch procedure developed by Mononobe and Ohtsu's group [6]. The first etch was silicone/49% HF, the second etch was a BHF solution with ratio of 10:1:1 (no silicone oil) for 2 hours, followed by a third etch with four-fold diluted HF (no silicone oil) for 90 seconds. Fibers were washed with deionized H_2O, acetone and methanol after first etch, then with deionized H_2O after second and third etch.

To examine the overall shape of the etched fibers, Nikon OptiHot 88 microscope at the University of Utah Microfabrication Lab was used. The microscope had objective lens at 5X, 10X, 20X, and 60X. However, due to optical diffraction limit, the resolution of the TOF was poor, and SEM was used to acquire better images for tip size and cone angle measurements. To prepare the fibers for SEM, a layer of conductive metal was deposited on the TOF. All etched TOFs were gold (Au) coated by Denton Discovery Sputtering System (Denton Vacuum, Moorestown, NJ), with chamber pressure of 2E-6 mTorr, and voltage of 50W. Some of the TOF samples were sputtered with chromium (Cr) at 100W for 8 seconds prior to the gold layer, and the thickness of that chromium layer was ~8nm. This was done because general literature suggested a Cr layer would provide better adhesion [7]. Both Au and Cr target were pre-sputtered for 1 minute. The Au sputtering rate was 20nm Au per minute, some samples were sputtered for 1.5 minutes, and others for 2 minutes to yield approximately 30 ± 10 and 40 ± 10 nm thickness of gold, respectively.

Two SEM facilities were used to obtain SEM images. A Hitachi S3000-N SEM, operated in high vacuum mode only, and low beam voltage (5kV – 10kV) was used to obtain optimum resolution of a 2 μm scale. Another SEM machine Philips XL30 ESEM FEG (FEI, Hillsboro, Oregon), with low vacuum mode available, allowed better image resolution (200 nm scale) than high vacuum mode. The setting for low vacuum mode was 1-1.4 Torr in vacuum, secondary electron mode with beam power of 10kV.

Prior to Au coating, laser power transmissions through TOFs with varying cone angles were measured using argon laser at 515 nm (BeamLok2580, Spectral Physics, Irvine, CA). The non-tapered end of the fiber was inserted into a chuck holder (FPH-J, Newport, Irvine, CA), which was placed in a fiber coupler with X, Y and Z adjustments (F-915 Newport, Irvine, CA) along with a 10X objective lens (M-10, Newport, Irvine, CA) to focus the incoming light. The input power was 40 ± 1 mW, but only19 ± 1 mW is measured after the light passed through a filter, before it entered the objective lens. A power meter measured the light at the end of the TOF, and transmission % was calculated based on the 19 ± 1 mW entering the apparatus.

Results

Optical and SEM images were acquired for the TOF samples as shown below, the optical images of Figure 1 were taken at different time points from the same fiber to calculate the etch rate. Based on the measurements, the etch rate of the cladding in silicone/49% HF was 3.54 ± 0.17 μm/min, which is in good agreement with other literature reporting [5].

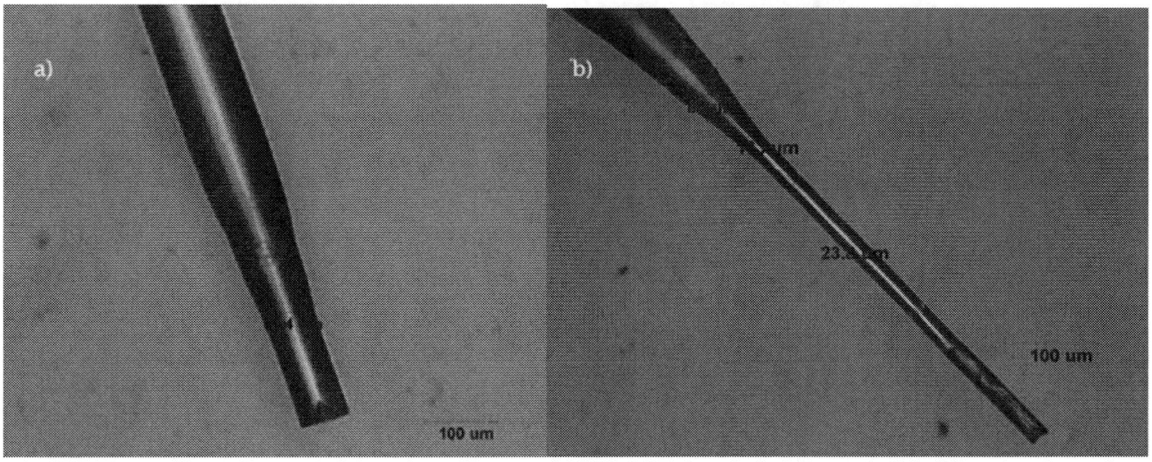

Figure 1. TOF profile after etching in silicone/49% HF for (a) 15 and (b) 30 minutes. Images taken by Nikon Optihot 88 microscope at 10X lens. The etch rate of the cladding was 3.54 ± 0.17 μm/min.

For Variation I, the SEM pictures of the TOF at different etching time intervals are shown in Figure 2. To better characterize the relationship between etching time and cone angle, data are plotted in Figure 3. The graph indicated that cone angle increased as etching time increased, but that trend changed after the 90-minute time point, when the cone angles showed little or no change.

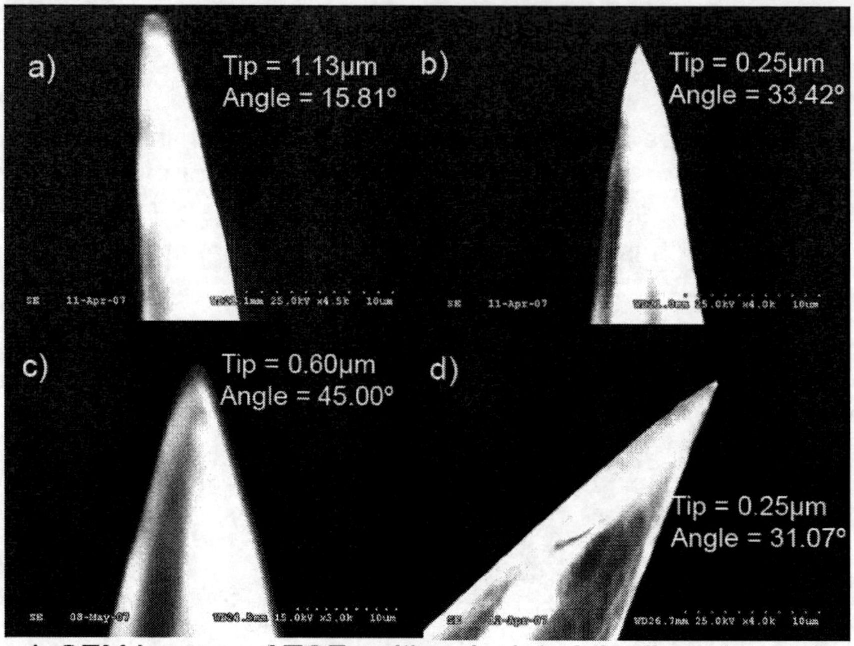

Figure 2. Variation I: SEM images of TOFs with calculated tip diameter and cone angle dimensions at different etching time. TOFs etched in silicone oil/HF solution for (a) 40-minute (b) 60-minute (c) 90-minute (d) 120-minute. Cone angles at the tip increased with etching time until after 90 minutes.

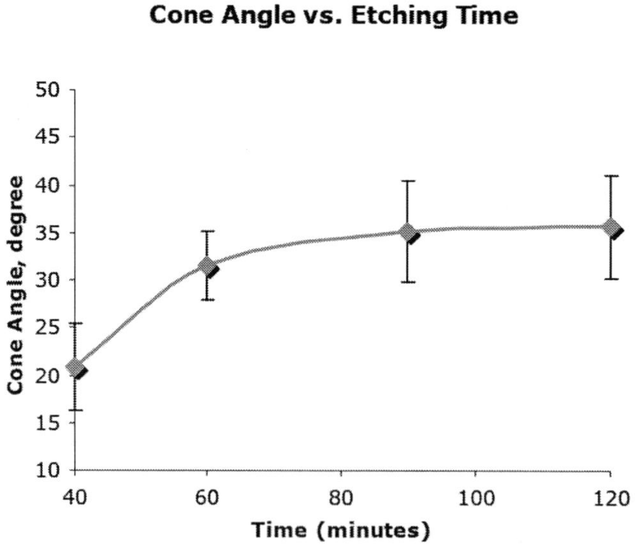

Figure 3. Plot of cone angle vs. etching time for Variation I. The rate of change in cone angle is minimal after 90 minutes.

In Variation II, TOFs were second etched with four-fold diluted HF solution (see Figure 4). Although the smallest tip diameter obtained was 117 nm, the dimensions of tip diameter and cone angles were not reproducible between batches. From previous experimental measurements, the etch rate of the diluted HF at the fiber core was estimated to be 150 nm/min.

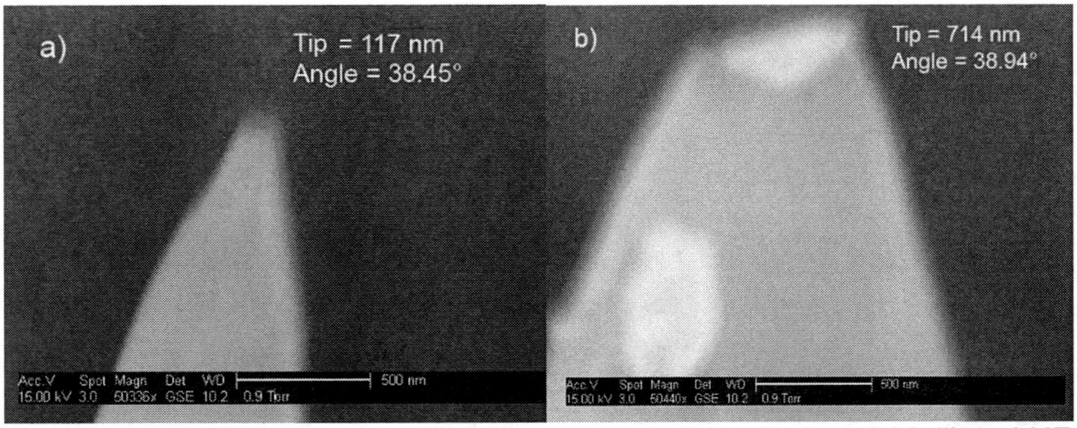

Figure 4. Variation II: SEM image of TOFs after second etch with four-fold diluted HF. (a) 50-minute first etch, 40-minute second etch (b) 60-minute first etch, 40-minute second etch.

In the procedure of variation III, the first etch of the TOFs was for 50 minutes, the second etch time intervals were varied (see Figure 5). The 60-minute second etch had produced a smaller tip size of 70 nm, but also smaller cone angle 20°.

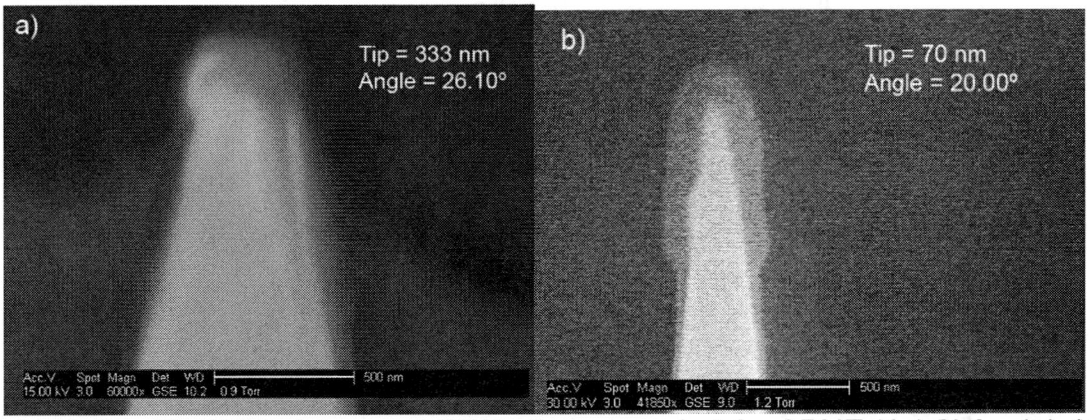

Figure 5. Variation III: SEM pictures of TOF after second etch of BHF (4:1:3) for (a) 40 minutes (b) 60 minutes. The halo feature around the tip was carbon artifact due to the environment in the SEM chamber.

SEM images of Variation IV showed that 50-minute interval for first etch yielded TOF with smaller tip compared to the 60-minute interval (see Figure 6). In their triple-etch procedure, Mononobe and Ohtsu's stated 50 minutes for first etch.

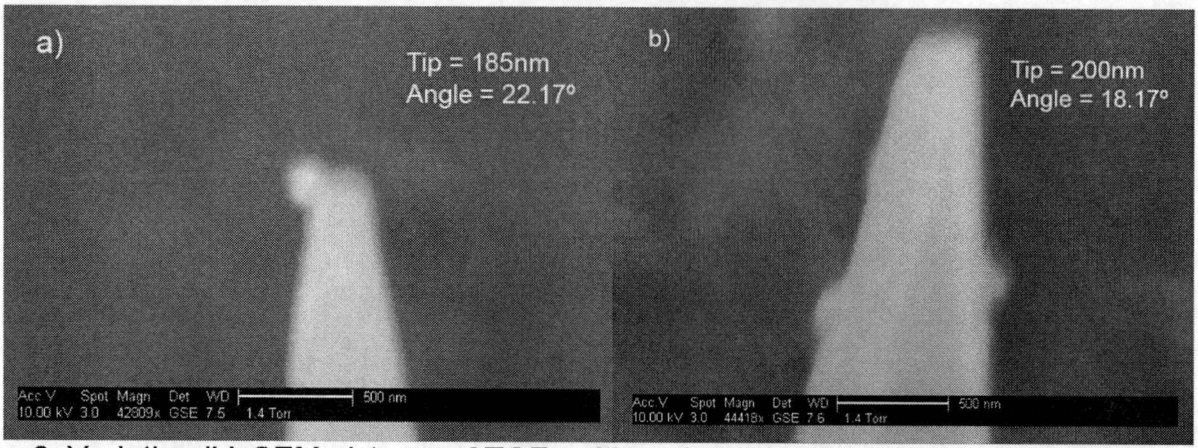

Figure 6. Variation IV: SEM pictures of TOFs with varied time points for first etch at (a) 50-minute (b) 60-minute. The cluster on the side of the fiber was likely due to contamination during the sputtering process.

Data collected from argon laser transmissions is plotted as shown in Figure 7 as a function of cone angles. Note that the tip sizes were not the same for the different TOFs due to the immense difficulty in producing the same tip size for tips with different cone angles. The tip size parameter could have introduced the variation within the data points. Overall, there appeared to be a general increase in power transmission as cone angle increases.

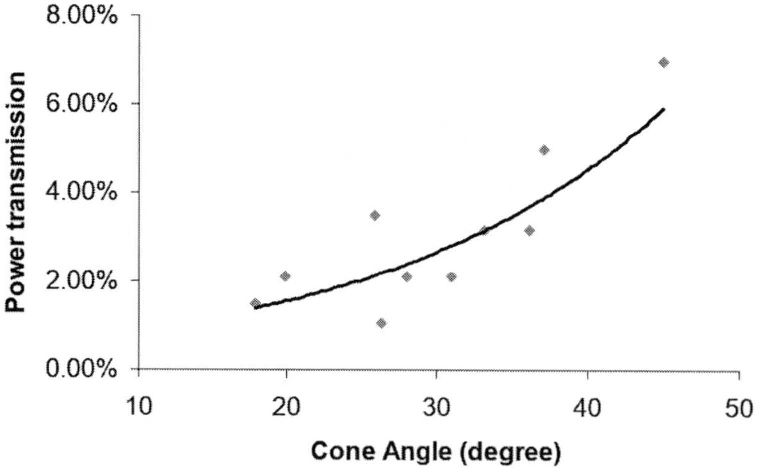

Figure 7. Measurement of power transmissions of TOFs with different cone angles.

Discussion:

The chemical reaction of HF etching was described by Stockle et al. [8] as follows:

$$SiO_2 + 4HF \rightarrow SiF_4 + 2H_2O,$$
$$3SiF_4 + 2H_2O \rightarrow SiO_2 + 2H_2SiF_6$$

because the cladding of the fiber is less dense than the core of the fiber, any etching will proceed faster at the cladding and slower once the core is exposed. The microscope images from Figure 1 indicated that the etch rate of the cladding in silicone/49% HF etch was 3.54 ± 0.17 μm/min. When the HF etching starts, the meniscus height at the interphase will decrease because the surface tension is decreasing with the cladding surface area being etched away. This mechanism is the driving force for the tapered tip geometry.

The SEM data indicated that Variation I and II were not effective in fabricating TOF with tip diameter below 100 nm. Variation IV yielded tip diameter of about 200 nm, and Variation III yielded a tip size below 100 nm. Reproducible results were obtained from Variation III to consistently fabricate TOF that has tip diameter of 80 ± 10 nm and cone angle of ~20°.

The laser power measurements confirmed that cone angle is an important parameter in determining transmission efficiency of the TOFs. The data suggested that bigger cone angle yields relatively higher power transmission than smaller cone angle. For the smaller cone angle fiber, the light traveling through the core reached a cut-off radius sooner and some of the light was reflected back, hence it resulted in very low transmission through the tip.

The application of BHF appeared to be crucial in etching the tip diameter to small sizes after the first etching. It is very likely that the etch rates of 49% HF and the diluted HF (12.25%) were too aggressive and over etched the fine tip of the TOF, thus resulted in tips ranging from 250 nm to 1000 nm. And since BHF is buffered with NH_4F, the etching was much slower and thus better controlled. Mononobe and Ohtsu reported etch rate of BHF (ratio of 10:1:1) at the fiber core to be 100 nm/min [9]. It can also be concluded that the ratio of BHF effect the tip size in that lower concentration of NH_4F (in ratio 4:1:3) was more beneficial in producing smaller and smoother tip than higher concentration of NH_4F (in ratio 10:1:1), and high concentration of NH_4F typically resulted in TOF with bumpy surface. Comparing the cone angle dimensions from the SEM images, the cone angles tended to be smaller after the second etch, and did not exceed 30°. This suggested that it is harder to fabricate small tip diameter with big cone angle, because there is only so little to etch away before reducing the cone angle. It is likely that there is a trade-off between cone angle and desired tip diameter.

Reference

[1] A. Rasmussen, V. Deckert, **New dimension in nano-imaging: breaking through the diffraction limit with scanning near-field optical microscopy,** *Anal Bioanal Chem (2005)* 381: 165-172.

[2] P. Burgos, Z.Lu, A. Ianoul, et al., Near-field scanning optical microscopy probes: a comparison of pulled and double-etched bent NSOM probes for fluorescence imaging of biological samples. *Journal of Microscopy,*Vol. 211, Pt 1 2003, 37-47.

[3] BB. Akhremitchev, Y. Sun, L. Stebounova, GC. **Walker, Monolayer-sensitive infrared imaging of DNA stripes using apertureless near-field microscopy,** *Langmuir*, **2002**, *18(14)*, 5325-5328.

[4] K. Maruyama, H. Ohkawa, S Ogawa et al., **Fabrication and characterization of a nanometer-sized optical fiber electrode based on selective chemical etching for scanning electrochemical/optical microscopy.** *Anal Chem.* 2006, 78, 1904-1912.

[5] Pak Kin Wong, Tza-Huei Wang, Chih-Ming Ho, "Optical Fiber Tip Fabricated by Surface Tension Controlled Etching", Hilton Head 2002 Solid State Sensor, Actuator and Microsystems Workshop, pp. 94-97, June 2-6, 2002, South Carolina, USA.

[6] S. Mononobe, R.Uma Maheswari, M. Ohtsu, **Fabrication of a pencil-shaped fiber probe with a nanometric protrusion from a metal film for near-field optical microscopy.** Optics Express (1997), Vol. 1, Issue 8, pp. 229-233,

[7] H. Frey, F. Keilmann, A. Kriele, R. Guckenberger, **Enhancing the resolution of scanning near-field optical microscopy by a metal tip grown on an aperture probe.** *Appl. Phys. Lett.* Vol. 81, 26 (2002).

[8] R. Stockle, Fokas, Deckert et al., **High-quality near-field optical probes by tube etching.** *Appl. Phys. Lett..* 75, 160 (1990)

[9] S. Mononobe, M. Ohtsu. **Fabrication of a pencil-shaped fiber probe for near-field optics by selective chemical etching**. *J. Lightwave Technol.* **1996**, *14,* 2231

426c Nanowire-Based Electrochemical Recordings of Catecholamines

Christina Randall[1], Chih-Sheng Chiang[1], Kartik Murari[1], Zhiyong Gu[2], Nitish Thakor[1], and **David H. Gracias**[3]. (1) Department of Biomedical Engineering, Johns Hopkins University, Baltimore, MD 21218, (2) Department of Chemical Engineering, University of Massachusetts, One University Avenue, Lowell, MA 01854, (3) Department of Chemical and Biomolecular Engineering, Johns Hopkins University, 125 Maryland Hall, 3400 N Charles Street, Baltimore, MD 21218

Abstract—We describe electrochemical recording of catecholamines using gold and platinum nanowire (NW) electrodes with diameters of 30-200 nm and lengths of 1-5 μ m. The NW width (30 nm) falls within the range of the size of the synaptic cleft, which is 20-50 nm in length. The electrochemical devices were fabricated by depositing NWs on silica substrates followed by photolithographic patterning of the contact pad for electrical contact. Special care was taken to expose only the NW surface to the catecholamine solution during calibration, by covering the entire surface of the contact pad with photoresist using a second photolithographic step. Calibration curves were generated in both beaker experiments and microfluidic channels to assess the ability of the nanoelectrodes to detect dopamine (DA) and serotonin (5-HT) at predicted synaptic concentrations. Using the NW electrodes, we were able to detect DA and 5-HT in mM concentrations. This sensitivity demonstrates the utility of the device in measuring predicted synaptic concentrations of DA (approximately 1.6 mM) and for investigating the changes in 5-HT concentration during epileptic seizures. The NW recording devices also exhibit calibration parameters comparable to current carbon fiber electrodes (linearity of 0.995). The small electrode size and the ability to record predicted synaptic concentrations point to the possibility of using the NW electrode device to do neural recording from a synaptic cleft. BACKGROUND IN the late 70's, Adams used voltammetry to demonstrate that catecholamines could be oxidized in-vivo by pulsing a constant voltage current through an electrode inserted into brain tissue [1]. The resulting oxidation current was recorded using a potentiostat and was found to be proportional to the extracellular concentration of the monoamine near the tip of the recording electrode. This measurement capability is important to record changes in the extracellular concentration of these monoamines because they have been implicated in several mental illnesses, including attention deficity hyperactivity disorder [2], schizophrenia [3] and epilepsy [4]. Since then, carbon fiber electrodes have become the current standard and are approximately 10 μ m in diameter causing minimal damage to the neural tissue during recording. Nevertheless, even 10 μm diameter electrodes can record extracellular neurotransmitter concentrations from only a cluster of neurons. Our goal was to develop a nanosensor that was small enough to fit into the synaptic cleft between two neurons, as a first step to enable single neuron recording. Hence we developed a nanosensor with a diameter in the size range of the synaptic cleft (20-50 nm). CURRENT RESULTS Nanowires ranging from 30 to 200 nm were fabricated by electrodeposition of the appropriate metal in anodized nanoporous membranes as described elsewhere [5]. After electrodeposition, the NWs were released from the membrane by dissolution in sodium hydroxide. The NWs were then rinsed repeatedly with water and ethanol and then stored in ethanol. The NW electrodes were fabricated as shown in Fig. 1. Briefly, the NWs were dispersed on a silica substrate. Using photolithography, a contact pad consisting of a 50 nm chromium adhesion layer and 200 nm copper was aligned with one end partially covering the electrode. Since the surface area of the contact pad was much larger than that of the NW, and the metal of the pad would contribute to the current measured, an insulation layer of photoresist was patterned to completely cover the contact pad while leaving the NW exposed (Fig. 2). Calibration curves were conducted in phosphate buffered saline using a Keithley model 6430 sourcemeter (Keithley Inc., Cleveland, OH) in a two terminal potentiostatic configuration (Fig. 3a). In all experiments, the voltage across the terminals was fixed at 0.55V for DA and 0.45 V for 5-HT and the current was monitored for analysis. For each electrode, data from a minimum of three successive aliquot additions, and two consecutive repeat experiments were then used to generate a linear regression fit (calibration curve, Fig. 3b). The slope of the resulting calibration curves represent the sensitivity of the electrode to DA concentrations. It should be noted that the data points for each aliquot in the calibration curve were determined from the 20-80% range average of the signal to avoid skewing of the data by the initial spike in current that occurs upon the addition of the DA. From the calibration curves, the sensitivity, linearity and their correlation to NW surface area were then determined. The 200 nm Au (sensitivity: 2.37 μA/mM, R2 = 0.992) and Pt (sensitivity: 1.29 μA/mM, R2 = 0.955) electrodes were comparable. The statistical differences observed can be attributed to photolithographic alignment, surface charge, presence of edge plane sites and oxidation of the nanowire surface 17. The 30 nm Au nanosensor however, had a sensitivity of 0.27 μA/mM and had increased linearity over both of the 200 nm nanosensors (R2 = 0.996). REFERENCES [1] R. N. Adams "Probing brain chemistry with electroanalytical techniques," Analytical Chemistry, vol. 48, pp. 1126A-1138A, 1976. [2] M. V. Solanto "Attention-deficit/hyperactivity disorder: clinical features". In: psychostimulant drugs and ADHD: basic and clinical neuroscience, pps. 3-30, (Eds. M Solanto, AFT Arnsten, FX Castellanos), Oxford, Oxford University Press pp. 3-30, 2001. [3] L. Linner, C. Wiker, M. L. Wackenberg, M. Schalling, and T. H. Svensson. "Noradrenaline reuptake inhibition enhances the antipsychotic-like effect of raclopride and potentiates D2-blockage-induced dopamine release in the medial prefrontal cortex of the rat," Neuropsychopharma, vol. 27, pp. 691-698, 2002. [4] W. C. Ziai, D. L. Sherman, and M. A. Mirski. "Target specific catecholamine elevation induced by anticonvulsant thalamic deep brain stimulation," Epilepsia, vol.. 46, pp. .878-881, 2005. [5] H. Ye, Z. Gu, T. Yu and D. H. Gracias "Integrating nanowires with substrates using directed assembly and nanoscale soldering," IEEE Trans Nanotech., vol. 5(1), pp. 1-6, 2006.

426b Glass Nanopore Electrode And Membrane Sensors

Chett J. Boxley[1], James T. Steppan[1], Henry S. White[2], and Bo Zhang[3]. (1) Ceramatec, 2425 S 900 W, Salt Lake City, UT 84119, (2) Department of Chemistry University of Utah, University of Utah, 300 South 1400 East, Salt Lake City, UT 84112, (3) Chemistry, Pennsylvania State University, University Park, PA 16802

Recent reports in molecular biology and analytical chemistry have led to a new class of molecular sensors based on transport in biological and artificial pores of nanometer dimensions. In this presentation, we describe a bench-top method for reproducibly fabricating glass nanopore (GNP) electrodes and membranes with pore orifice radii between 2 and 500 nm. Prototype sensors using the GNP electrodes and membranes have been constructed for applications in nanoparticle counting, stochastic single-molecule detection, and in ion selective electrochemistry.

GNP electrodes and membranes are fabricated by sealing an electrochemically sharpened Au or Pt micro-wire into a glass capillary, followed by polishing the glass until a nanometer-sized metal disk is exposed. The key to the success of this methodology is the use of a high-input impedance metal-oxide semiconductor field effect transistor (MOSFET)-based circuit to monitor the radius of the disk electrode during polishing. Proper biasing of the MOSFET circuit, based on consideration of the polishing circuit impedance, enables the routine, bench-top fabrication of Pt nanodisk electrodes of pre-selected radius as small as ~10 nm. In subsequent steps, the fabrication of a GNP electrode is accomplished by etching the metal nanodisk electrode to create a pore in glass, with the remaining metal disk comprising the pore base. Complete removal of the wire yields a glass nanopore membrane, in which a conical shaped pore is embedded in a thin (~50 micrometer) GNP membrane.

The nanopore electrode and membrane can be modified by numerous chemical means including: covalent attachment of molecules to the surfaces; filling the pore volume with polymers and hydrogels; and by deposition of lipid bilayer/proteins across the pore orifice. These modifications impart molecular selectivity and high sensitivity (single molecule) in designing sensors for applications in homeland defense, DNA sequencing, environmental monitoring, and particle counting.

426e Patterned Electrodes For Thickness Shear Mode Quartz Resonators To Achieve Uniform Mass Sensitivity Distribution

Anthony Richardson[1], Venkat R. Bhethanabotla[1], and Allan L. Smith[2]. (1) Department of Chemical Engineering, University of South Florida, 4202 East Fowler Ave., ENB118, Tampa, FL 33620, (2) Masscal Scientific Instruments, 12565 Research Parkway, Suite 300, Orlando, FL 32826

Development of a sensor instrument that is responsive to minute mass loadings on a nanoscale, while exhibiting a mass sensitivity profile that is independent of material placement on the sensor platform, would enable nanogram measurements of a broad range of materials, a necessity in chemical and biological research. Mass sensitivity of current commercial analytical balances is limited to the microgram scale which is inadequate as scientific and technological research continues to progress towards further minute magnitudes. Fortunately, significant research into the utilization of thickness shear mode (TSM) quartz resonators as nanoscale mass balances is well documented1. However, a substantial shortcoming of past research is the establishment of a flat radial sensitivity distribution across the sensor platform. Non-uniformity of mass sensitivity on the TSM resonator surface prohibits the utilization of this technology in the measurement of absolute mass. To achieve a uniform distribution requires the design and testing of various electrode patterns. Presented in this study are the radial sensitivity distributions of TSM devices having different electrode geometries. In conducting the experiments, the electrodes were deposited on blank AT-cut quartz wafers and the wafers placed inside a G1 mass/heat flow instrument developed by Masscal Scientific Instruments. This instrument permits the extraction of the frequency signal of the crystal, either bare or solvent droplet loaded, and the associated thermal power which enter as parameters in the mathematical models describing mass sensitivity. Additionally, a computer simulation of the mass sensitivity was conducted for each electrode using finite element method (FEM) analysis to corroborate the experimental results, and to discover the best possible design.

References:

[1] Josse, F.; Lee, Y.; Martin, S. J.; Cernosek, R. W. Analysis of the radial dependence of mass sensitivity for modified-electrode quartz crystal resonators. Analytical Chemistry 1998, 70, 237-247.

[2] Smith, A. Gravimetric analysis of the non-volatile residue from an evaporated droplet, using the quartz crystal microbalance/heat conduction calorimeter. J. ASTM Intl. 2006, 3, 1-5.

426f Application Of CuO_x-CeO_2 Catalysts As Selective Sensor Substrates For Catalytic Detection Of CO In H_2 Fuel

Christopher S. Polster and Chelsey D. Baertsch. Chemical Engineering, Purdue University, Forney Hall of Chemical Engineering, 480 Stadium Mall Drive, West Lafayette, IN 47906

Proton exchange membrane fuel cells (PEMFCs) using H_2 fuel are a promising source of portable power for electronic devices and transportation. However, as CO is a common poison for PEMFC electrodes and also a common contaminant in H_2, it is critical to be able to detect and quantify CO contamination at low levels in concentrated H_2 fuel. On-line CO detection in portable devices requires sensors with small size, low cost, low power consumption, and high sensitivity. These requirements have been met by microelectromechanical system (MEMS) signal transducers. However, all current microsensor technologies use materials that are inherently unselective to CO, give rise to detrimental false positive responses in the presence of H_2, and require arrays of sensors for gas identification. A new sensor paradigm will be presented utilizing an intrinsically selective catalytic reaction coupled with micro thermal transduction. In this sensing mechanism, a substrate selectively and exothermically oxidizes CO to CO_2, causing a measurable temperature rise that can be related to CO concentration.

CuO_x-CeO_2 catalysts have been shown to oxidize CO to CO_2 with up to 100% CO oxidation selectivity in process streams containing ~10,000 ppm CO or more in H_2 excess. It is critical to maximize selectivity towards CO in order to avoid false positive responses resulting from H_2 oxidation. CO oxidation selectivity (relative to H_2 oxidation) must be nearly 100%; the potential sensing range of the catalyst at a particular operating temperature is determined by the concentration at which selectivity drops from this value. The lower limits at which CuO_x-CeO_2 maintains this high selectivity were found to be 200 and 500 ppm at operating temperatures of 333 and 353 K, respectively. Thus, CuO_x-CeO_2 catalysts are a promising sensing substrate for ppm level detection of CO in H_2 fuels by reaction calorimetry. The high selectivity of this catalyst is shown to be a product of both high kinetic selectivity and high surface coverage of CO. The loss of 100% selectivity at decreasing CO concentration is attributed to a lower driving force for CO adsorption. Integration of this catalyst with MEMS sensor technologies will provide new capabilities in portable sensors for the selective quantitative detection of CO in H_2 while avoiding false positive responses.

481a Electrochemical Multiphase Microreactor As Fast, Selective, And Portable Chemical Sensor Of Trace Organophosphorus Vapors

Chelsea N. Monty[1], Ilwhan Oh[1], and Rich Masel[2]. (1) Department of Chemical and Biomolecular Engineering, University of Illinois at Urbana-Champaign, 294 RAL, MC-712, 600 S. Mathews Ave., Urbana, IL 61801, (2) Chemical Engineering, University Of Illinois, 600 S Mathews, Urbana, IL 61801

Due to an increase in demand, the detection of organophosphorus compounds (OPs) has become extremely important. Conventional methods for detection of OPs are gas chromatography/mass spectroscopy (GC/MS) and ion mobility spectrometry (IMS). GC/MS and IMS are not ideal detection methods for trace levels of OPs due to their high false positive rates, high cost of production, and low portability.

The objective of this work is 1) to create a micro-scale gas-liquid interface using microchannels and nanoporous membranes and 2) to adapt the oxime chemistry into the microchannel sensor in order to build an electrochemical sensor which selectively detects the trace (part-per-billion or lower) gas-phase organophosphorus compounds. This paper describes testing of oxime sensors with organophosphorus compound D. First, macro-scale experiments were conducted to determine the response of the oxime solution to compound D. Oxime chemistry was then adapted into two microchannel sensor designs: a passive and an active sensor. Both sensors were exposed to gaseous compound D and the response was recorded.

481b Integrated Electrical Sensor Arrays In Microfluidic Networks

Matthew C. Cole and Paul J. A. Kenis. Chemical and Biomolecular Engineering, University of Illinois at Urbana-Champaign, 600 S. Mathews Ave., Urbana, IL 61801

As microfluidic devices continue to decrease in size and increase in complexity, the ability to monitor the passage of material throughout them becomes ever more important. Efficient sensing of the position of discrete fluidic elements in microfluidics may eventually lead to the creation of viable routing and scheduling systems for the microscale. Many current microfluidic sensing techniques commonly employ optical methods based on differences in the refractive index of fluids. These techniques, though accurate, can often be complicated and difficult to incorporate with the rest of the microfluidic fabrication process. Optical sensors for microfluidics are also very difficult to scale down as microchannels approach the nanometer regime. A promising alternative to optical sensors involves the use of electrical elements (either resistors, capacitors, or conduction gaps) embedded in microfluidic devices. Electrical sensors can be less demanding in space, less expensive, and can offer a greater number of sensors per unit area than their optical counterparts.

We have used thin film resistive heaters and coplanar capacitive sensors to demonstrate the detection of discrete plugs of alternating fluids in a microfluidic networks. Resistive sensors detect position by applying a small, constant potential across a resistor and continuously monitoring the current through it. Changes in current are caused by changes in the thermal conductivity of the fluid surrounding the resistor, and thereby signify the presence of a new fluid plug at the position of the resistor. In the capacitive case, a potential is applied across the gap between two coplanar electrodes, and the current is monitored. Changes in the fluid bridging the gap result in a sudden and sharp induced current spike in the output, and indicates the detection a fluid plug.

To fully realize the goal of a microfluidic control system, these individual electrical sensors must be integrated into a large array spanning an entire microfluidic network. A major problem associated with this is that as the number of sensors increases, so too does the number of electrical leads necessary to connect the sensors with external monitoring equipment. This rapid growth can make the fabrication and implementation of the sensor array exceedingly difficult. The creation of a large array of sensors that also minimizes the number of electrical leads necessary is therefore desirable. The present work proposes a solution to this problem by introducing a multiplexing approach that allows an array of $m \times n$ sensors to be controlled by only $m+n+1$ electrical leads. Each sensing element is connected to two electrical leads, and each electrical lead is connected to multiple sensing elements. When a change in fluid occurs at some sensor (m,n), a change in the monitored output value is displayed in the trace of both lead m and lead n at the same time. This allows one to pinpoint exactly where a fluid element is in a microfluidic system with only $m+n+1$ leads.

As a proof of the multiplexing concept, 4×4 arrays of resistive and coplanar capacitive sensors were used to monitor the passage of discrete fluid plugs throughout a microfluidic network. This presentation will focus on the fabrication, design rules, and applicability of these sensors to different types of fluid transport detection in various microfluidic architectures.

481c Hydrodynamic Focusing Studies In A Microreactor By Electrochemical Techniques

Abhishek G. Deshpande, Sinead M Matthews, Kamran Yunus, Adrian C Fisher, and Nigel K. H. Slater. Department of Chemical Engineering, University of Cambridge, New Museums Site, Pembroke Street, CB2 3RA, Cambridge, United Kingdom

A microelectrochemical reactor (MECR) was developed, fabricated and tested for electrochemical detection and analysis of potassium ferricyanide. The application and advantage of hydrodyamic focusing by using a background electrolyte has been demonstrated. High conversions were observed with this new technique and the cell was further developed using computational modelling to improve the conversions and study the effect of lead in length and focusing ratios in order to minimise the diffusion effects in a microchannel over the electrode. It was demonstrated that by increasing the lead in length inter stream diffusion increases leading to less amount of material being addressed at the electrode and hence lower conversions were obtained.

481d Reversible Single Walled Carbon Nanotube Electronic Sensor Arrays Integrated With Micro-Gc Column

Chang Young Lee and Michael Strano. Chemical Engineering, Massachusetts Institute of Technology, 77 Massachusetts Avenue, Building 66-366, Cambridge, MA 02139

A wide range of analytes adsorb irreversibly to the surfaces of single walled carbon nanotube electronic networks typically used as sensors or thin-film transistors, although the mechanism remains poorly understood. Using thionyl chloride as a model electron-withdrawing adsorbate, we show that reversible adsorption sites can be created on the nanotube array via functionalization with amine-terminated molecules of $pK_a < 8.5$. The reversible response enables extraction of the characteristic free energy of adsorption for each type of amine. A nanotube network comprised of single, largely unbundled nanotubes, near the electronic percolation threshold is required for the effective conversion to a reversibly binding array. In-situ Raman spectroscopy confirms that adsorption takes place on the amine coating and does not distinguish between SWNT by electronic type. By screening 11 types of amine-functionalization, we show that analyte adsorption is largely affected by the basicity of surface groups. A mechanism of reduced binding energy by surface chemistry is proposed based upon XPS and molecular potential calculations. Parts-per-trillion level sensitivity is demonstrated by creating a higher adsorption site density with a polymer amine, such as polyethyleneimine (PEI). The ability to systematically tune molecular adsorption to maximize both analyte sensitivity and reversible response has applications to the detection of trace analytes in real time from a micro-fluidic chromatographic column. We demonstrate, for the first time, that a pulse as small as $\sim 10^9$ molecules of dimethyl methylphosphonate (DMPP), a nerve agent simulant, is reversibly detected at the end of a micro-fabricated gas chromatographic (μGC) column coupled to the functionalized SWNT array.

481e Improved Love-Wave Biosensor Through Advanced Interdigitated Transducer Design

Stefan Cular[1], Subramanian Sankaranarayanan[1], Venkat R. Bhethanabotla[1], and Darren W. Branch[2]. (1) Department of Chemical Engineering, University of South Florida, 4202 East Fowler Ave., ENB 118, Tampa, FL 33620, (2) Biosensors and Nanomaterials Department, Sandia National Laboratories, Albuquerque, NM 87185

We present an experimental and theoretical study of surface acoustic wave (SAW) interdigitated transducer (IDT) designs used to generate Love-waves for biosensor applications. Three IDT designs were created on 36°YX $LiTaO_3$ selected for its high electro-mechanical coupling and shear horizontal (SH) wave generation. These designs were selected based on a literature review of IDT designs of improved electrical characteristics within communication applications. It was shown that the sensitivity of the SAW biosensor can be significantly increased through the addition of a waveguide to generate Love-mode waves instead of SH. For the Love-mode generation to occur: 1) the waveguide material should have a lower acoustic velocity than the substrate, and 2) the waveguide material should be of lower density than the substrate. In previous studies, many materials have been shown to be efficient waveguides. Some researchers have even developed multilayer schemes to further enhance the sensors sensitivity. The most used materials are silicon dioxide, poly(methyl methacrylate), polyimide, and polystyrene. In this study, we have utilized polystyrene as the waveguide material to allow for comparison with previous studies. Parameters studied in this work were the attenuation and velocity changes of the acoustic wave due to the addition of varying thicknesses of polystyrene. Experimental data was collected using a network analyzer with time domain measurement capabilities, and theoretical results were calculated using a 3D finite element model (FEM). Two kinds of FEM analysis are carried out along each of the three delay lines: (1) An impulse input of 100 V over 1 ns is applied to study the frequency response of the device for the various IDT designs and (2) AC analysis with a 5 V peak-peak input and approx. 100 MHz frequency to study the wave propagation characteristics. Our results indicate the significant differences between IDT designs for sensor applications.

481f Method For Measuring Thickness Of Dielectric Films Using Microdielectric Fringe-Effect Sensors

Prashant Tathireddy, University of Utah, 50 S. Central Campus Dr., Salt Lake City, UT 84112, Yunn-Hong Choi, Central R&D Institute, Samsung Electro-Mechanics Fundamental Technology Center, South Korea, and Mikhail Skliar, Chemical Engineering, University of Utah, 50 S. Central Campus Dr., Salt Lake City, UT 84112.

A method for noninvasive thickness measurements of dielectric films using fringe-effect (FE) sensors is developed and experimentally validated. The fringing electrical field, created by electrodes microfabricated at the film substrate, depends on the film thickness and dielectric permittivity of the film under test (FUT). The unknown film thickness is estimated by matching the theoretical prediction of thickness-dependent sensor admittance with the measured value. In the case of FE sensors with spatially periodic, interdigitated electrode (IDE) configuration, the admittance prediction is simplified, which allows for the real-time measurements of changing thickness. The developed method can be used to continuously measure the changing dielectric permittivity of the FUT material, which makes it possible to determine the thickness of films of changing dielectric properties, caused by chemical or other transformations. The application of the developed method is demonstrated experimentally by measuring the thickness of silicon nitride film deposited in several increments on the quartz substrate of the IDE sensor. In the expected range of sensor sensitivity, the results show an excellent agreement with the independent thickness measurements.

Reference:

Y. Choi, P. Tathireddy, and M. Skliar, Anal. Chem. 78:3242-3248, 2006.

523a A Biochip for Rapid and Sensitive Detection of Multiple Cancer Markers Simultaneously

Edgar D. Goluch[1], Savka Stoeva[2], Kashan A. Shaikh[1], Sandra S. Szegedi[1], Jae-Seung Lee[2], Thomas N. Chiesl[3], Annelise E. Barron[3], Chad A. Mirkin[2], and Chang Liu[1]. (1) Micro and Nanotechnology Laboratory, University of Illinois at Urbana-Champaign, 208 N. Wright St. MNTL, Urbana, IL 61801, (2) Department of Chemistry and Institute of Nanotechnology, Northwestern University, 2145 Sheridan Road, Evanston, IL 60208, (3) Chemical and Biological Engineering, Northwestern University, 2145 Sheridan Road, Evanston, IL 60208

It is well established that early cancer detection increases patient survival rates. The most obvious approach is to employ cancer markers for disease detection, but thus far, no single cancer marker has been demonstrated to conclusively predict a disease state. To address this problem we are developing diagnostic tools that are capable of multiplexed detection in a single sample. Toward this goal, we have designed a fully integrated, chip-based system for carrying out simultaneous, high-sensitivity detection of two protein cancer markers. Previously, we reported chip-based detection of a single protein target [1, 2, 3]. However, until now we have not demonstrated multiplexed detection, which presents significant new challenges. Here, we have addressed the many challenges associated with achieving a chip-based multiplexed assay. We have improved the microfluidic chip design, identified the proper chip coating to minimize nonspecific binding of probes, and advanced the identification of multiplexed biobarcode oligonucleotide strands [4]. Compared to the equivalent bench-top assay [5], the chip-based system decreases the reagent volumes, shortens the assay time (to less than three hours for the entire assay), and achieves comparable sensitivity. The shortened assay time assay time is attributed to the fact that the chip-based system forces the target molecules and probes into close proximity, leading to faster capture and decreased overall assay time. Our system also eliminates the need for several pieces of equipment used in the bench-top assay including incubators, vortexes, and centrifuges.

[1] E.D. Goluch, et al., "A Bio-Barcode Assay for On-Chip Attomolar-Sensitivity Protein Detection", Lab-on-a-Chip, vol. 6, pp.1293-1299, 2006. [2] K.A. Shaikh, et al., "A Modular Microfluidic Architecture for Integrated Biochemical Analysis", PNAS, vol. 102, pp. 9745-9750, 2005. [3] E.D. Goluch, et al., "A Microfluidic Device Using Nanoparticle-Based Bio-Barcodes for Ultra-Sensitive Detection of Protein", in MicroTAS'04, Malmo, Sweden, September 26-30, 2004, pp. 342-344. [4] E.D. Goluch, et al., "Chip-Based High-Sensitivity Detection of Multiple Disease Biomarkers", in MEMS'06, Istanbul, Turkey, January 22-26, 2006, pp. 478-481. [5] S.I. Stoeva, et al., "Multiplexed Detection of Protein Cancer Markers with Biobarcoded Nanoparticle Probes", J. Am. Chem. Soc., vol. 128, pp. 8378-8379, 2006.

523b Biosensor Incorporating Cell Barrier Architectures For Screening Cancer

Gargi Ghosh, Department of Chemical and Materials Engineering, University of Kentucky, Lexington, KY 40506, Kimberly Anderson, Chemical Engineering, University of Kentucky, 3900 Steele Rd, Versailles, KY 40383, and Leonidas Bachas, Chemistry, University of Kentucky, Lexington, KY 40506.

According to American Cancer Society 559,650 deaths are expected from cancer making it the second most death causing disease after heart diseases. In most cases, detection of cancer at the early stages is not possible as it involves complicated invasive procedures. In the absence of a quick screening tool, many individuals are diagnosed in the latter stages of the disease thereby reducing the chance of successful therapy. It is a known fact that early screening of cancer can reduce cancer death by nearly 50%. Screening can not only detect cancer at early stages but can also identify an individual showing an inclination towards developing cancer. Thus at present there is an urgent need for a quick screening tool for cancer. Research efforts have demonstrated that higher concentrations of cytokines like VEGF, HGF, bFGF, TNF-alpha, etc are released in the blood of cancer patients as compared to healthy individuals. These angiogenic molecules are also known to modulate the permeability of endothelial cells. A cell based biosensor has been developed in our laboratory. This sensor consists of a confluent monolayer of human umbilical vein endothelial cells (HUVECs) seeded on the surface of an asymmetric cellulose triacetate (CTA) membrane of ion selective electrode (ISE). Earlier studies demonstrated that upon confluent monolayer formation, complete inhibition of ion transport occurs across the membrane resulting in reduced response. Exposing the sensor to permeability modifying agents results in a concentration dependant response from the sensor. The present study focused on applying the biosensor as a screening tool for cancer. For this purpose, the biosensor was exposed to the above mentioned cytokines, individually and in combinations, at concentrations observed in the blood of cancer patients and compared to the responses obtained when exposing the sensor to the cytokines at concentrations found in healthy patients. The study demonstrates the ability of the sensor to differentiate between these concentrations of cytokines when measuring both individual and combinations. Currently, multiplexed ELISA is capable of simultaneously detecting a number of individual growth factors. However, since during the early stages of cancer, individual concentrations of these factors are low, detection using ELISA poses a difficult problem. Since, the biosensor can measure not only individual cytokines but also combinations of them; we hypothesize that this biosensor will be able to detect cancer at the early stages. Present studies are directed towards exposing the biosensor to the serum of healthy individuals and cancer patients and measuring the responses. These results will also be discussed.

523c Development Of Label-Free Nanopattern-Enhanced Biosensors For Food Safety Monitoring And Early Cancer Diagnostics

Shaoyi Jiang, Chemical Engineering, University of Washington, Benson Hall, Box 351750, Seattle, WA 98195

In this work, our recent efforts in the development of label-free surface plasmon resonance (SPR) sensors for food safety monitoring and early cancer diagnostics will be reviewed. For instrumentation, a nanopattern-enhanced SPR platform has been developed and then integrated into an SPR imaging system with contrast polarization. Nanopatterns are generated with electron beam lithography (EBL), which allows for control over all aspects of the nanofeatures including diameter, separation distance, and height. For surface chemistry, an array of proteins (antibodies, single-chain Fvs, or antigens) has been developed via spotting on a dual functional carboxybetaine-based coating, which is not only highly resistant to protein adsorption even from 100% blood plasma, but also has abundant functional groups for protein immobilization. For food safety applications, we have been working on the simultaneous and quantitative detections of bacterial pathogens, viruses, and marine toxins in complex media including juices, milks, and shell fish extracts in collaboration with researchers from the U.S. FDA. For early cancer diagnostics, we have been working on the screening of biomarkers (either antigens or auto-antibodies) from clinical samples for ovarian and breast cancers in collaboration with researchers from the Fred Hutchinson Cancer Research Center.

523d Fractal Analysis of Binding and Dissociation Kinetics of Cancer Markers on Biosensor Surfaces

A. Sadana, University of Mississippi, Rebel Drive, University, MS 38655-1848

A fractal analysis is presented for the binding and the dissociation of different cancer markers on biosensor surfaces. The data analyzed include putrescine in solution to PDDA/APTES/MWCNT/PuO-modified GCE (glassy carbon electrode (Rochette et al., 2005), and vascular endothelial growth factor (VEGF) in solution to the soluble form of the VEGF receptor (sFlt-1 or VEGF-1) immobilized on a sensor chip surface (Barleon et al., 1996). Single- and dual-fractal models were used to fit the data. Values of the binding and the dissociation rate coefficient(s), affinity values, and the fractal dimensions were obtained from the regression analysis provided by Corel Quattro Pro 8.0. The binding rate coefficient and the affinity values are sensitive to the degree of heterogeneity present on the sensor chip surface. Predictive equations are developed for the binding rate coefficient as a function of the heterogeneity present on the biosensor chip surface. For example, the binding kinetics of different concentrations of putrescine (25-100 micromolar) in solution to a PDDA poly (diallyldimethylammonium) chloride/APTES (3-aminopropyltrietoxy-silane)/MWCNT (multi-walled carbon nanotube/PuO (putrescine oxidase)-modified GCE (glassy carbon electrode) in 0.05 M phosphate buffer were analyzed. The binding rate coefficient, k of putrescine in solution to plasma from a tumor-disposed mouse or a healthy mouse exhibits an order of dependence between second and two and one-half order on the fractal dimension, Df or the degree of heterogeneity that exists on the biosensor surface. Similarly, for the binding of VEGF in the 5-2250 ng/mL concentration range in solution to its receptor (sFlt-1) immobilized on a biosensor chip surface the binding rate coefficient, k exhibits an order of dependence between one and one-half and second on the VEGF concentration in solution. The analysis presented provides physical insights into these cancer-biomarker-receptor reactions occurring on the different biosensor surfaces.

523e Nano-Metal Particle Reagent For Highly Sensitive, Real-Time, Fluorophore Mediated Biosensor

Jianting Wang, Bin Hong, and Kyung A. Kang. Chemical Engineering, University of Louisville, Ernst Hall 106, Speed School of Engineering, University of Louisville, Louisville, KY 40292

Fluorophore mediated, immuno-biosensors have been demonstrated to be effective disease detection tools. To obtain accurate diagnosis, successful sensor must provide high specificity and sensitivity. The sensitivity of fluorophore mediated biosensor highly depends on the intensity of the fluorescence from the target molecules. Therefore, a method to enhance fluorescence emission from a fluorophore can improve the sensitivity of the sensor. Nanometal particle (NMP) placed at an appropriate distance can effectively enhance the fluorescence by transferring the free elections of the fluorophore, normally used for self-quenching, to the strong surface plasmon polariton fields of the nanoparticles. We have found this enhancement depend on the metal type of NMP, NMP size, distance between a fluorophore and an NMP, and quantum yield of a fluorophore. We also found some biocompatible solvents could enhance the fluorescence significantly, possibly via effective dipole coupling between the fluorophore and the solvent molecule, resulting in an increase in the energy gap between the excited and the ground state. Conformational change of the protein complexes in solvent can be another contribution to the enhancement by reducing the distance between fluorophore and sensor surface. To maximize the fluorescence enhancement, we developed nanogold particle reagents (NGPRs) by combining nanogold particles and biocompatible solvents. These NGPRs were tested and optimized in four-cardiac marker fiber-optic biosensing system and found to be able to increase the signal as much as 10 times and accurately quantify cardiac markers at tens of picomolar level. Authors acknowledge the financial support from Kentucky Science and Engineering Foundation (KSEF-148-502-03-55) for fluorescence enhancement studies and National Science Foundation (BES-0330075) for cardiac marker biosensing.

523f A Three-Dimensional Microfluidic System Integrated with Surface Plasmon Resonance Microscopy for Immunoassays

Jianping Liu[1], Mark A. Eddings[2], Bruce K. Gale[3], and Jennifer S. Shumaker-Parry[1]. (1) Department of Chemistry, University of Utah, 315 S 1400 E, Rm 2020, Salt Lake City, UT 84112, (2) Department of Bioengineering, University of Utah, 50 S Central Campus Dr, Salt Lake City, UT 84112, (3) Department of Mechanical Engineering, University of Utah, 50 S Central Campus Dr., Salt Lake City, UT 84112

We demonstrate the combination of a high-throughput microfluidic device with surface plasmon resonance (SPR) microscopy for quantitative, in situ antibody immobilization. The microfluidic device provides 48 separate flow channels that can be used simultaneously for antibody immobilization and subsequent antibody-antigen interaction analysis. The microfluidic system is made from an elastomeric material and the design is based on a standard 96-well plate. Each well in the system is linked to a second well through a pair of channels that meet to form a single V-shaped microfluidic channel. When the microfluidics system is placed at the sensing surface of the SPR microscope, each channel forms a sensing area of ~ 400 x 400 μm. In the current configuration, 48 microfluidic channels form 48 separate sensing regions. One major advantage of the microfluidic system is that the antibody immobilization process is performed under humid conditions to help retain the activity of the antibody. In addition, SPR microscopy provides quantitative characterization of the immobilization process in real time. We are developing high-throughput immmunoassays based on this integrated system. The presented methods could be extended to probe other biomolecule interactions in a label-free, high-throughput and quantitative way.

523g Hexagonal Saw Interleukin-6 Biosensor

Stefan Cular[1], Venkat R. Bhethanabotla[1], Darren W. Branch[2], and Joel A. Strom[1]. (1) Department of Chemical Engineering, University of South Florida, 4202 East Fowler Ave., ENB 118, Tampa, FL 33620, (2) Biosensors and Nanomaterials Department, Sandia National Laboratories, Albuquerque, NM 87185

We present the development and application of a hexagonal surface acoustic wave (SAW) biosensor for Interleukin-6 at levels required for physiological relevance. Interleukin-6 is a proinflammatory cytokine involved in the body's pathophysiologic response to injury due to various causes, e.g. trauma, burns, sepsis, and disease. We have optimized the sensor configuration to provide a substantially improved response in comparison with traditional SAW sensors. The hexagonal SAW biosensor was fabricated in 36° YX $LiTaO_3$ to generate shear horizontal (SH) waves using the high electro-mechanical conversion efficiency of this substrate material. The biosensors were coated with an optimally thick polystyrene waveguide enabling Love-wave generation. The generated Love-waves were shown to improve sensitivity 40-fold. Results from the human Interleukin-6 analyte biosensor will be presented. The biosensor was fabricated using an anti-human Interleukin-6 biosensor film physically adsorbed to the polystyrene waveguide. The resulting data were collected using a homebuilt sensor circuit connected to a multimeter and personal computer for data acquisition, and were verified using fluorescent microscopy.

576a Acetylcholinesterase-Based Electrochemical Multiphase Microreactor for Detection of Trace Organophosphorus Vapors

Chelsea N. Monty[1], Ilwhan Oh[1], and Rich Masel[2]. (1) Department of Chemical and Biomolecular Engineering, University of Illinois at Urbana-Champaign, 294 RAL, MC-712, 600 S. Mathews Ave., Urbana, IL 61801, (2) Chemical Engineering, University Of Illinois, 600 S Mathews, Urbana, IL 61801

Due to an increase in demand, the detection of organophosphorus compounds (OPs) has become extremely important. Conventional methods for detection of OPs are gas chromatography/mass spectroscopy (GC/MS) and ion mobility spectrometry (IMS). GC/MS and IMS are not ideal detection methods for trace levels of OPs due to their high false positive rates, high cost of production, and low portability.

The hydrolysis of acetylcholine (ACh) is the principal step that terminates intercellular pathways. Organophosphorus compounds with a similar structure to ACh block this hydrolysis through inhibition of acetylcholinesterase (AChE). The objective of this work is to fabricate a micro-scale device which utilizes a gas-liquid interface to more rapidly detect the inhibition of AChE through chronoamperometry. The miniaturized sensor exhibits fast kinetics, is highly selective, allows for detection of vapor compounds in real time, and has parts per trillion detection limits. This paper describes the optimization of the AChE biosensor. The effect of liquid and gas flow rates, enzyme and substrate concentrations, and buffer pH on enzyme inhibition is studied. Also, recovery of the inhibited enzyme by antidotes, such as pralidoxime, is investigated.

576b A Theoretical And Experimental Study Of Multilayered Bioelectronic Interfaces Containing Dehydrogenase Enzymes

Brian L. Hassler[1], Claire Vieille[2], Scott Calabrese Barton[3], and Robert M. Worden[3]. (1) Chemical Engineering and Materials Science, Michigan State University, 2527 Engineering Building, East Lansing, MI 48824-1226, (2) Biochemistry and Molecular Biology, Michigan State University, 410 Biochemistry Building, Michigan State University, East Lansing, MI 48824, (3) Michigan State University, Department of Chemical Engineering and Materials science, Michigan state university, East Lansing, MI 48824-1226

Bioelectronic interfaces that establish electrical communication between redox enzymes and electrodes have potential applications as biosensors, biocatalytic reactors, and biological fuel cells. We have used layer-by-layer self assembly to fabricate bioelectronic interfaces containing the polyelectrolytes polyethylenimine (PEI) and polyacrylic acid (PAA) functionalized with dehydrogenase enzymes and their cofactors. The interfaces are able to achieve mediated electron transfer between the enzyme and the underlying electrode. The effect of the number of layers of PEI, mannitol dehydrogenase (MtDH), PAA, and toluidine blue O (TBO) on the performance of the bioelectronic interface was studied using cyclic voltammetry, chronoamperometry, constant potential amperometry, and electrochemical impedance spectroscopy. The current generated by the interface increased initially with the number of layers, and then reached a limiting value. As the number of layers increased, the calculated surface coverage increases 1-fold with each additional layer, but the apparent electron transfer coefficient and turnover rate decreased. A mathematical model of the multilayered bioelectronic interface has been developed and used to interpret the experimental results.

576c Diamond Microneedle Electrodes for Neurochemical Detection

Heidi B. Martin, Case Western Reserve University, A.W. Smith Building, 10900 Euclid Ave., Cleveland, OH 44106

To be added

576d Micropatterned Fluid Lipid Bilayers Created Using A Continuous Flow Microspotter For Multi-Analyte Assays

Kathryn A. Smith[1], Bruce K. Gale[2], and John C. Conboy[1]. (1) Department of Chemistry, University of Utah, 315 S. 1400 E. RM. 2020, Salt Lake City, UT 84112, (2) Department of Mechanical Engineering, University of Utah, 50 S. Central Campus Drive, Salt Lake City, UT 84112

There is an increasing interest in the development of microarray based assays for high-throughput analysis and detection of biomolecules such as DNA and proteins. We have developed a new method for creating multianalyte planar supported lipid bilayer (PSLB) arrays using a 3D continuous flow microfluidic system (CFM™). PSLBs are an attractive platform for performing biological assays due to their high resistance to nonspecific biomolecule adsorption and non-fouling nature. PSLBs also serve as good models of biological membranes, allowing a diverse assortment of biological and chemical interactions to be explored in a controlled manner in vitro. The CFM™ allows for an addressable microchannel to be formed, in which an array of fluid bilayers can be patterned onto a glass substrate. The system's capability of performing multianalyte assays has potential applications in many fields such as biosensing, drug discovery, proteomics and clinical diagnostics.

A Bead-Based Microfluidic Immunosensor for Small Molecule Detection

Dwayne Vickers[1] and Tamara Floyd-Smith[2]
[1]Mechanical Engineering Department and [2]Chemical Engineering Department, Tuskegee University, Tuskegee, AL

Abstract

A bead-based microfluidic immunosensor has been developed to detect small molecules. Proof of concept is demonstrated by sensing trinitrotoluene (TNT). The microfluidic device has a one entrance, one exit configuration, with a region that allows the beads to be immobilized. The flow pattern and pressure drop across the beads for various bead sizes was investigated to determine the relationship between the pressure drop and the bed length and bead size. Next, biotinylated TNT antibodies were attached to avidin coated beads. Detection of TNT was accomplished via a displacement assay and downstream fluorescence spectroscopy.

Introduction

Recent trends in science have shifted towards the development of miniaturized total analysis (lab-on-a-chip, microfluidic, etc.) systems that have the following potential advantages over conventional macroscopic instrumentation: (i) reduced sample and reagent consumption, (ii) shorter analysis time, (iii) higher sensitivity, (iv) portability, (v) disposability and low production cost [1].

This work focuses on using a bead-based lab-on-chip device that detects small molecules, specifically, trinitrotoluene (TNT). The military sector requires devices that monitor environmental levels of energetic materials such as TNT. Compounds like TNT have been introduced into soils and water supplies on and surrounding military bases in the US and abroad by production, testing, storage and disposal of weapons. Hence, in the recent years environmental monitoring and remediation has been increasing, causing the military sector to require portable devices that can provide on-site, real-time, semi-quantitative or quantitative measurements of contaminant levels [2].

For this work, avidin-coated beads loaded in a microchannel captured the biotinylated antibody and facilitated TNT detection using a displacement assay and fluorescence detection. The antibody for the target analyte, TNT, is biotinylated so when the antibody is passed through the channel, it binds irreversibly to the avidin-coated beads. The antibody-to-bead binding concept is illustrated in Figure 1.

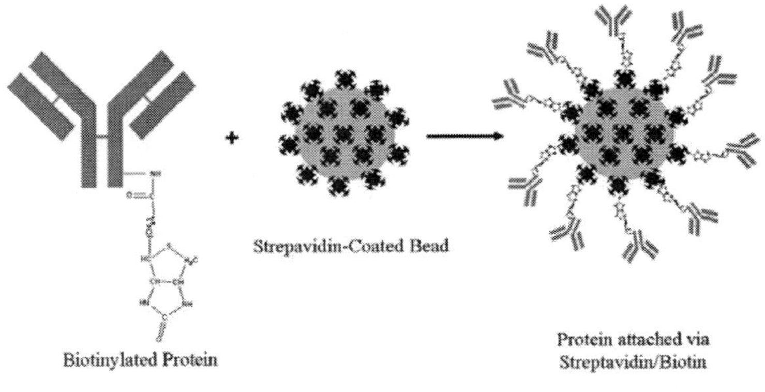

Figure 1: Illustration of Biotin-Avidin Binding

There are several types of assays that can be run using the immunosensor that we have fabricated. Avidin and biotin make highly specific and stable complexes; hence, they are highly used in biological analysis techniques such as immunoassays.

The variety of assay types used is based on the different types of analytes and samples to which they are most beneficial. There are two fundamental classifications of the techniques that exist, heterogeneous and homogeneous. Heterogeneous immunoassays, involve the antibody (or in some cases antigen) being immobilized on a solid support, while homogeneous assays involves the complete assay being performed in the solution phase. Heterogeneous immunoassays have the benefit of high surface area/volume ratio in the microchannel, while homogeneous assays have the ability to multiplex and give very fast electrophoretic separations.

The classifications of immunoassays can be further divided into competitive and noncompetitive. In competitive assays, the analyte competes in proportion to its concentration with the labeled analyte for a limited number of antibody binding sites. There is an increase in labeled analyte displaced as the analyte concentration increases, hence giving a decrease in signal if antibody-bound labeled analyte is detected or an increase in signal if the free labeled analyte is detected.

Noncompetitive assays, such as a sandwich immunoassay, have an increase in signal that is proportional to the analyte concentration. Heterogeneous sandwich immunoassays usually achieve separation by immobilizing a capture antibody on a solid phase and incubating to capture the analyte. Most heterogeneous immunoassay applications use the device or particles such as beads as the solid phase. An advantage of using the microchannel surface for antibody immobilization is the higher surface area to volume ratio than a standard well format. The antibodies are absorbed or patterned on the device surface by; direct adsorption on the channel walls, covalent attachment to reactive functional groups on the substrate, or micro-contact printing [3].

The use of micro beads as the solid support for the capture antibody is another type of heterogeneous immunoassay [4]. Beads dramatically increase the surface area in a small fluid volume, and dispersing them speeds up incubation times by reducing the diffusional distances. Preloading of the beads with the antibody and replacing with a fresh aliquot of beads between assays gives very rapid turnover times. There is also a high reproducible control of the antibody load being delivered to the channel from this strategy [5].

In our work a heterogeneous, competitive (displacement) immunoassay was used. We chose to use a displacement assay because it is appropriate for small target molecules, and it is also compatible with the detection scheme that we have chosen. Figure 2 illustrates the displacement assay where the analog is trinitrobenzene, a compound that will occupy the antibody sites until the preferred analyte, TNT, is in solution. The principle of this assay is that a fluorescently labeled compound (the analog) that cross reacts with the antibody is displaced from the antibody by the target analyte. The analog is fluorescently labeled so that the concentration of the target can be inferred using fluorescence spectroscopy from the amount of the analog that is displaced.

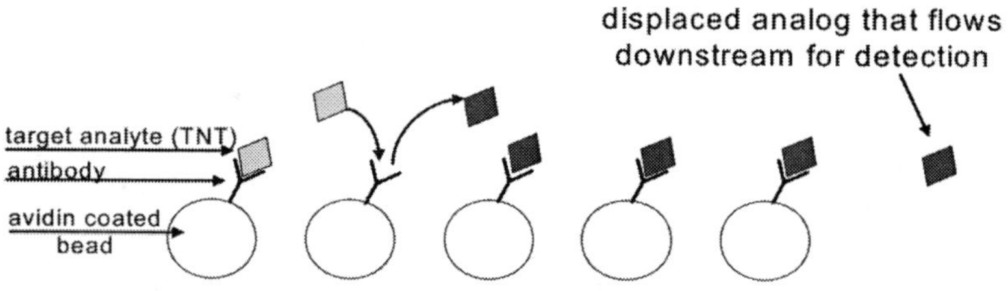

Figure 2: Illustration of the Displacement Assay

Experimental Procedure

Detection Methods

In this work, two different fluorescence detection methods were used. The first one was an offline spectrophotofluorometer, and the second was an online detection system shown in

Figure 3. Figure 3(a) shows that the detection system consists of a broad band light source, the immunosensor housing, a filter to transmit only a narrow wavelength of light, and a thermopile to detect the transmitted light. Figure 3(b) shows the syringe that holds the reagents and delivers them to the reactive sites. Figure 3(c) shows the detector in operation. The immunosensor is mounted so that the light passes through the filter, into the channel and finally to the detector.

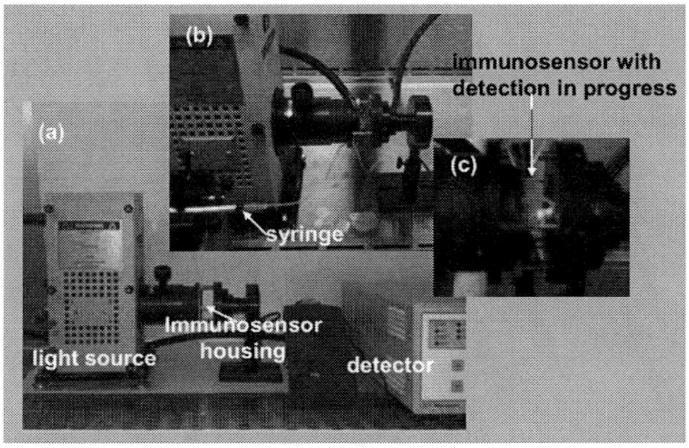

Figure 3: Picture of the On-line Detection System

Fabrication of Silicon Masters

The 5.3 mm wide and 2 cm long microchannels were fabricated in polydimethylsiloxane (PDMS) using soft lithography. The designs for the silicon masters were prepared using Macromedia Freehand MX software (Macromedia Inc.) and were printed on film using a 5080 dpi printer (Pageworks, Cambridge, MA). The films were subsequently used as photomasks during the lithography process. The silicon masters were prepared by spin coating SU-8, a negative epoxy-based photoresist (Microchem Corporation, SU-8 100, Newton, MA) onto silicon wafers. A layer of SU-8 100 was spun, baked on a hotplate for ~30 minutes, exposed through a transparency mask on a Karl Suss MA6/BA6 mask aligner and post-baked for ~10 minutes. The one-layer structure was then developed using SU-8 developer (Microchem Corp., Newton, MA). The fabrication process is illustrated in Figure 4, and the resulting silicon wafer with the immunosensor negative pattern is shown in Figure 5. Each four inch wafer has seven immunosensors as enumerated in Figure 5.

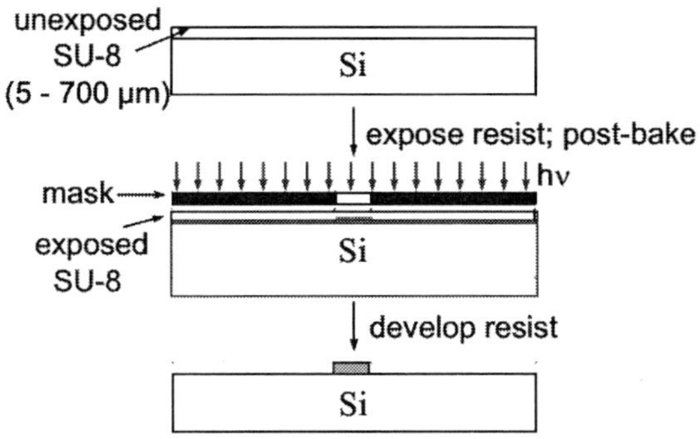

Figure 4: Steps for Fabrication of the Silicon Master

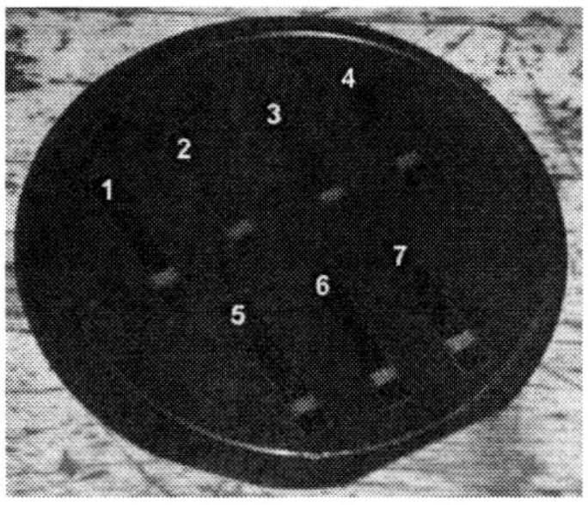

Figure 5: Silicon Wafer Used as a Master for the Immunosenor Molding

PDMS Patterning

After the photoresist-on-silicon master was fabricated, the surfaces were treated with tridecafluoro-1,1,2,2-tetrahydrooctyl trichlorosilane (Gelest Corporation, Morrisville, PA) in a vacuum chamber for 1 hour. The PDMS (Sylgard 184, Dow Corning, Corning, NY) was thoroughly mixed in a 10:1 mass ratio of silicone elastomer base to curing agent. This mixture was poured onto the silicon wafer and degassed in a vacuum chamber. Following the degassing process this mixture is cured in a convection oven for 2 hours at ~60°C. The PDMS was peeled away from the wafer, and cut to fit onto a microscope slide.

Bonding PDMS to Microscope Slide

The microscope slide was first cleaned with a 10 wt% potassium hydroxide/methanol mixture for ~30 minutes, washed with deionized water and dried thoroughly. The microscope slide and PDMS were placed in a plasma cleaner at full power (~ 12 MHz) for 1 minute. The surface of the slide and PDMS were contacted and a 1 psi pressure applied for ~5 minutes. Inlet and outlet holes were cut into the PDMS, and tubing, sealed with epoxy, was placed in the holes for fluid delivery. The completed immunosensor (less the beads) is shown in Figure 6.

Bead Preparation

Phosphate buffered saline with tween (PBST) with 10 mg/mL of Bovine Serum Albumin (BSA) was introduced into the channel at 0.15ml/min to passivate the channel and prevent non-specific binding to the PDMS. Next, the channel was loaded with 108 μm PMMA avidin-coated beads (Bangs Laboratories, Fisher, IN) as Figures 7(a) and 7(b) demonstrate. PBST was flowed into the channel at 0.5ml/nim to flush the system. Next, 10 μg /mL of biotinylated antibody (Naval Research Lab) in buffer was recirculated in the channel at 0.15ml/min. Finally, 10 μg /mL of the fluorescently labeled analog (TNB) was recirculated at a flowrate of 0.15ml/min, and the system was ready for detection of TNT.

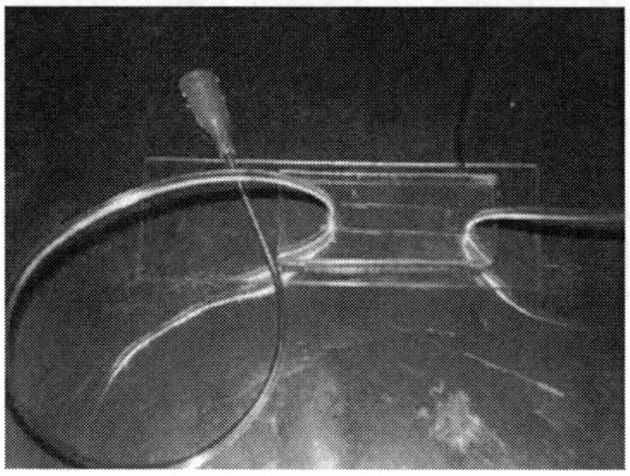

Figure 6: Immunosensor with Tubing Inserted

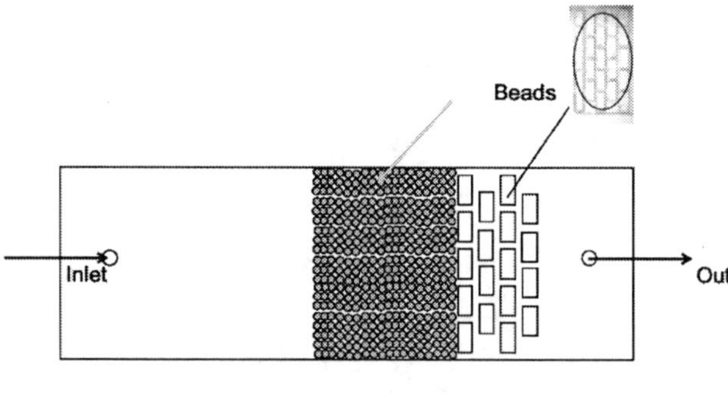

Figure 7a: Schematic of Immunosensor

Figure 7b: Optical Micrograph of Immunosensor

Pressure Drop Across Beads

PMMA beads (Bangs Laboratories, Fisher, IN) were pumped into the microchannel using a glycerol-water mixture with an equivalent density to that of the beads. The beads were loaded with bed lengths of 1 and 2 cm. Water, the working fluid used to measure the pressure drop, was pumped into the device, and a micro pressure sensor (Honeywell, Freeport, IL) measured the pressure drop across the beads. This micro pressure sensor was operated with the use of a multimeter and 10 voltage power supply. Figure 8 shows this experimental setup.

Results and Discussion

Pressure drop across the channel is an important characteristic of the system. Excessive pressure drops can suggest shearing within the system that could lead to cleaving of the bound reagents and also failure of the PDMS to glass bond for the immunosensor. Thus, pressure drop within the system was investigated and the results are shown in Figures 9 and 10.

Figure 8 : Experimental Setup for Pressure Drop Measurements

Bed lengths of 1.0 cm and 2.0 cm were investigated with flow rates ranging from 0.25 ml/min to 1.15 ml/min. The maximum pressure drop (Figure 9) observed was less than 4 psi for a 2.0 cm bedlength at 1.15 ml/min with a 116 μm bead diameter. Three different sizes of beads ranging from 116 μm to 250 μm were investigated with flow rates ranging from 0.25 ml/min to 1.15 ml/min. Figure 10 demonstrates that an increase in bead size results in a decrease in pressure drop. As the size of the beads increases in a fixed volume, there will be an increase in void space which results in the fluid passing through the beads more freely. The flow rate for the immunosensor is not expected to exceed 1.0 ml/min and the maximum bed length is 2.0 cm. A pressure drop on the order of atmospheric pressure is the limit of the PDMS to glass bond. Thus, from standpoint of the PDMS/glass bond, the system is operating well within its failure limits. Additionally, the pressure drop and flow velocities for this system are consistent with those of other immunosensors suggesting that the reagents will not be catastrophically impacted.

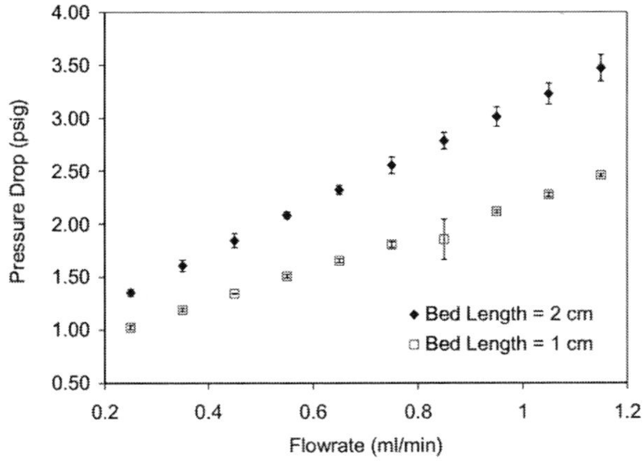

Figure 9: Varying Bed Length Pressure Drop Data

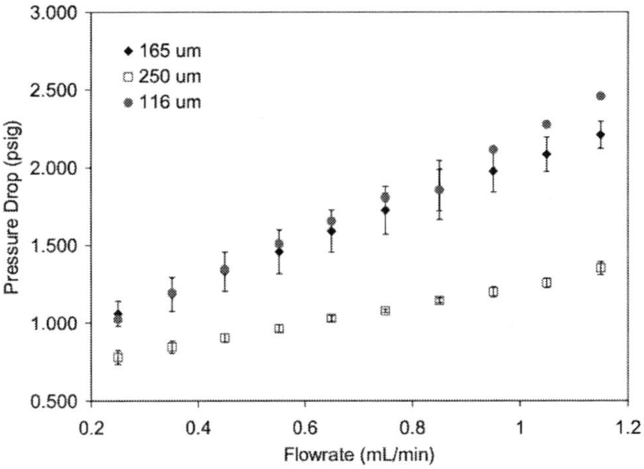

Figure 10: Varying Bead Size Pressure Drop Data

Prior to conducting an assay, the excitation and emission wavelengths for Cy5, the fluorophore, were determined. Figure 11, a plot of absorbance vs. wavelength, shows that the analog is maximally excited at 649 nm and the maximum emission wavelength is 666 nm. These results are also applicable at low concentrations.

Once the emission and excitation wavelengths were identified, a calibration curve (Figure 12) was constructed to correlate the concentration of TNB to the measured absorbance. This measurement is made to calibrate the fluorescence spectrophotometer prior to detecting the displaced TNT. The results indicate that the absorbance is linearly related to the concentration (μg/ml) by a coefficient of 0.26.

Finally, the displacement assay was conducted with known TNT samples (Figure 13). As expected, the absorbance increases with the concentration of TNT in the sample. The results also indicate that the displacement ratio is ~ 1 molecule of TNB: 3 molecules of TNT for a sample flow rate of 0.004 ml/min (~30 min residence time). Despite the long equilibration times, it is possible that the behavior is still flow rate dependent such that at higher flow rates or lower residence times fewer TNB molecules are displaced or vice versa. Displacement of fewer TNB molecules would lead to a sensor with a wider range of operation using fewer reagents at the expense of sensitivity.

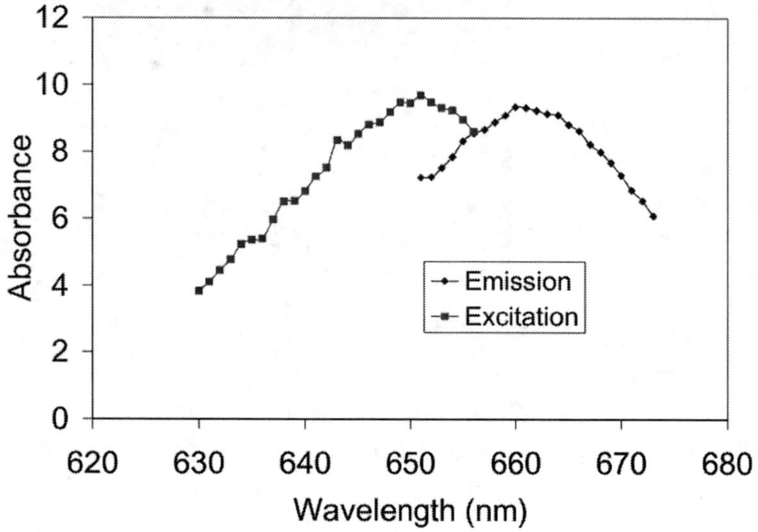

Figure 11: Excitation and Emission Wavelength for Cy5-labelled TNB

In principle, one might expect the TNB:TNT ratio to be stoichiometric. The absence of this stoichiometric relationship might suggest that the antibody sites were not initially saturated. To avoid this artifact, the antibody was exposed to the TNB for an extended period of time. Also, the intercept of the regression (-0.004) in Figure 13 corresponds to 0.05 μg/ml of TNT that would give a zero reading, suggesting that the antibody sites were saturated.

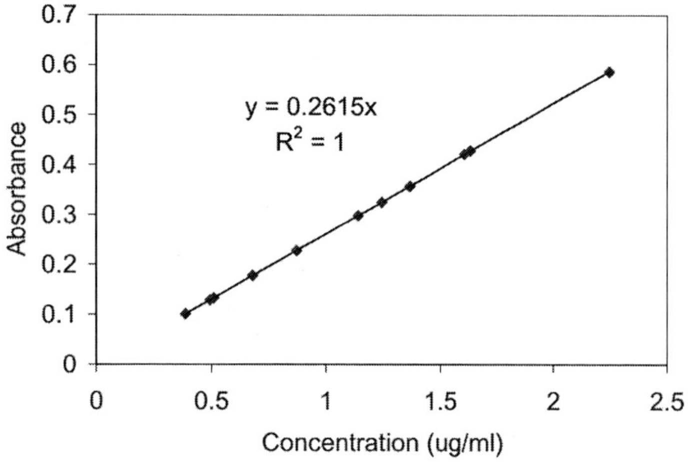

Figure 12: Calibration Curve for TNB Absorbance

Summary and Outlook

A bead-based microfluidic sensor for the detection of small molecules, specifically TNT, was demonstrated. In spite of limited data (due to limited reagents), the results look promising. However, more experiments are needed to investigate, in detail, various characteristics of the system such as whether the system behaves linearly and whether the sensitivity vs. range can be varied using residence time.

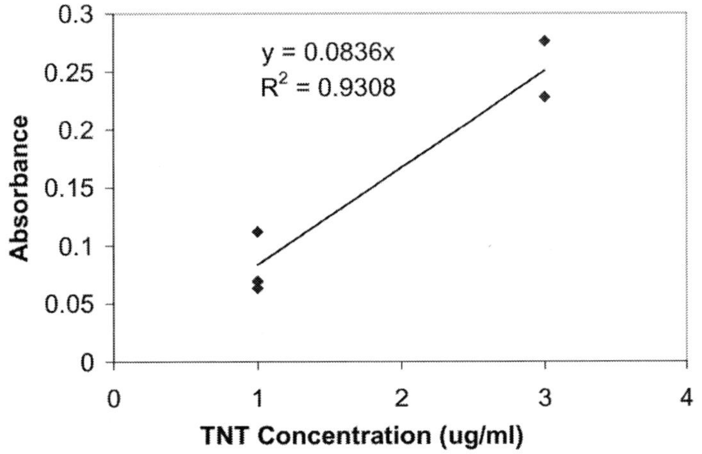

Figure 13: Demonstration of the Immunosensor for TNT Detection

Acknowledgements

The authors would like to acknowledge NSF (BES 0548788) for funding this research. We would also like to thank Paul Charles at the Naval Research Lab in Washington, DC for providing the TNT reagents and for helpful discussions.

References

1. M. Ali, R.Kirby, A. Goodey, *Anal. Chem.* **2003,** 75, 4732-4739.
2. D. Holt, P. Gauger, A. Kusterbeck, F. Ligler, *Biosensors and Bioelectronics* **2002**, 17, 95–103.
3. A. Kumar, G.M Whitesides, *Appl. Phys. Lett.* **1993**, 63 (14), 2002–2004.
4. C.A Wijayawardhana, H.B Halsall, *Anal. Chem.* **1999,** 399 (1–2), 3–11.
5. J.W. Choi, K.W. Oh, *Lab on a Chip 2.* **2002** (1), 27–30.

576f Fluid-Solid Interaction Study To Predict The Mechanism Of Removal Of Non-Specifically Bound Proteins In A Saw Biosensor

Subramanian Sankaranarayanan, University of South Florida, Dept. of Chemical Engg., ENB 118, 4202 E Fowler Avenue, Tampa, FL 33620, Venkat R. Bhethanabotla, Department of Chemical Engineering, University of South Florida, 4202 East Fowler Ave., ENB 118, Tampa, FL 33620, and Babu Joseph, Department of Chemical Engineering, University of South Florida, 4202 East Fowler Avenue, ENB 118, Tampa, FL 33620.

Background and Motivation: Fluid motion induced from high intensity sound waves is called acoustic streaming. Surface acoustic wave (SAW) devices (Fig. 1) used for biological species detection suffer from fouling that results from binding of non-specific protein molecules to the device surface. Non-specific binding dramatically reduces the sensitivity and selectivity of biosensors. The acoustic streaming phenomenon can be used to remove these non-specifically bound proteins to allow reuse of SAW devices (Fig. 2).

Methods: A finite element fluid solid interaction (FE-FSI) model of acoustic streaming phenomenon resulting from the interaction of surface acoustic waves (Rayleigh mode) with liquid loading is developed in this work. Parameters studied using the FE-FSI model include voltage intensity, device frequency, fluid viscosity and density. The transient solutions generated from the model were used to predict trends in acoustic streaming velocity. The generated streaming velocity fields are then utilized to predict the various removal forces experienced by the specific and non-specifically bound protein molecules located at various regions along the SAW biosensor delay path. Insights into the dominant mechanism responsible for efficient removal of non-specifically bound proteins can be obtained from the estimates of various forces predicted using perturbation methods and from the trends in the streaming velocity fields. For the sake of simplicity, the protein molecules are modeled as spherical shaped particles. The different SAW-induced forces which include linear forces such as added mass, drag, lift, and Basset forces and nonlinear ones due to radiation pressure, and drag exerted by acoustic streaming-are discussed and their magnitudes are evaluated for various particle sizes. The combined effect of the various interaction (adhesive and removal) forces is utilized to gain insights into the non-specifically bound protein removal mechanisms. The principal adhesion forces such as van der Waals forces and electrical double layer forces are used to model the interaction of the specific and non-specifically bound proteins to the biosensor surface. The forces responsible for removal of particles are mainly characterized as the direct SAW forces or the acoustic radiation forces as well as the lift and the drag forces that result from the mean velocity field in the fluid. The streaming velocity field (Fig. 3) computed using the FE-FSI model is utilized to estimate the fluid induced drag and lift forces for a range of operating conditions and fluid properties. Comparison of the various removal forces to the adhesive forces that bind the specific and non-specific proteins is used to estimate the removal efficiency along the SAW biosensor delay-line and optimize conditions which can enhance the same.

Results and Discussion: The results of our numerical simulation show that higher amplitude and high frequency surface acoustic waves can clean biosensor surfaces most effectively in media with fluid properties similar to those of water, i.e. lower viscosity and density, when (1) the SAW wavelength can be made comparable to the particle radius to promote effective acoustic-particle interaction; (2) the viscous boundary layer is thin and (3) both the non-linear acoustic streaming and radiation pressure forces exceed typical adhesion forces which is true for MHz frequencies. Based on the above analyses, possible mechanisms of non-specifically bound protein removal are discussed and interpreted in terms of the experimental observations of SAW biosensor surface cleaning.

Figures

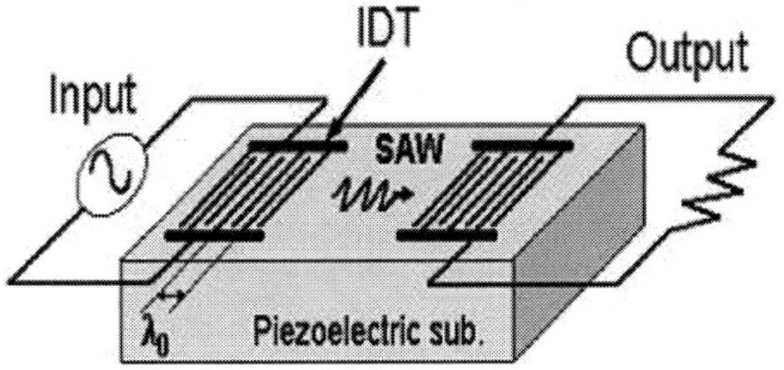

Fig. 1: Surface acoustic wave device used for chemical and bio-chemical sensing

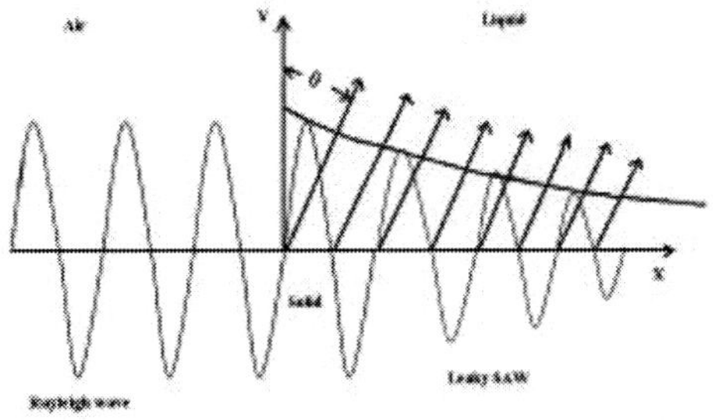

Fig. 2: Ultrasonic irradiation mechanism

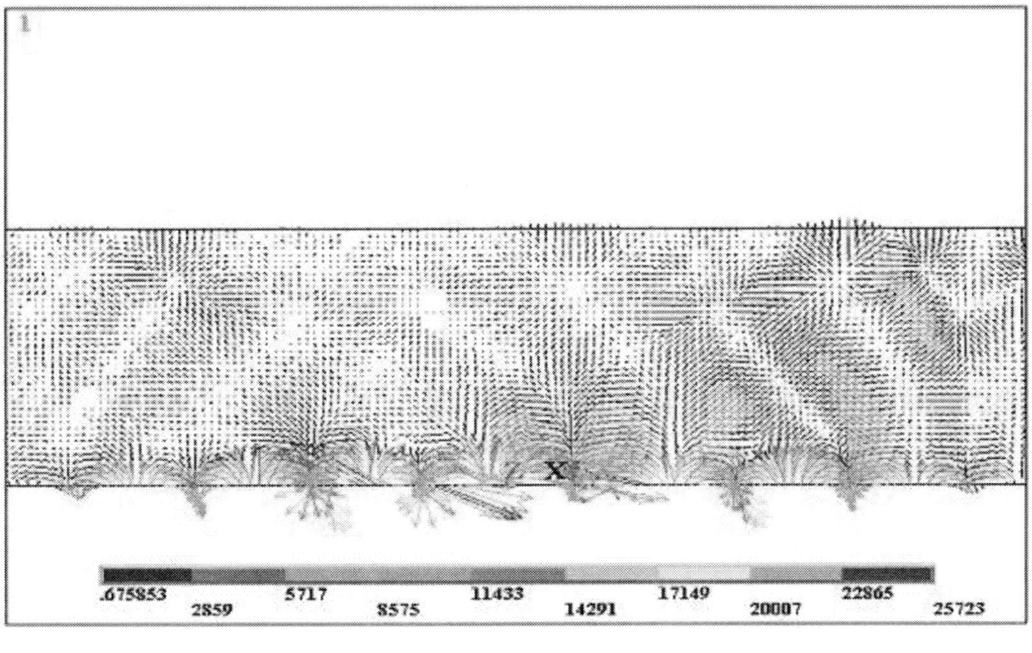

Fig. 3: Induced streaming patterns in the fluid domain for a 100 MHz frequency SAW device. The calculated velocities are in µm/s.

576g A Novel Computational Framework For The Optimal Design Of A Protein Biosensor That Recognizes A Known Set Of Ligands

Ho Ki Fung[1], Christodoulos A. Floudas[1], Addison D. Ault[2], and James R. Broach[2]. (1) Department of Chemical Engineering, Princeton University, Engineering Quadrangle, Olden Street, Princeton, NJ 08544, (2) Department of Molecular Biology, Princeton University, Princeton, NJ 08544

A biosensor is any ensemble of biological sensing elements which, upon interacting with an analyte or chemical species in an environment, is capable of marking that interaction by reacting measurably. The sensing elements can be biomolecules such as proteins, nucleic acids, or polysaccharides, viruses or bacteriophages, prokaryotes or eukaryotes, cellular organelles, cells, or organisms [1-2]. In this work, we propose a novel approach for designing a biosensor with protein receptors, whose ranges of stimuli can vary drastically. The problem can be stated as: given a blood sample in which a known set of drug ligands is present, and a set of protein receptors, we would like to design a biosensor with the minimal subset of receptors (i.e., optimal receptors) that is required for the optimal recognition of all ligands.

In order to address the biosensor design problem, we first devise a nonlinear programming (NLP) model to estimate the dissociation constants and efficacy values between each receptor-ligand pair. The model solves the parameter estimation problem based on the least squares minimization of the differences in the activation levels of any two receptors. The estimated values are then input into a novel mixed-integer nonlinear programming (MINLP) model [3] which both maximizes the sum of differences in the activation of any pair of receptors and minimizes the number of receptors needed to recognize all ligands. Constraints in the model include those relating the activation of each receptor to its dissociation constant and ligand concentrations, as well as others that guarantee concentration of each ligand is greater than the minimum of the dissociation constants for that ligand and all receptors selected to constitute the biosensor. The latter is to ensure that each ligand present in the blood sample will be recognized by the biosensor. In addition, we also impose upper and lower bounds on the ligand concentrations. Instead of generating only the optimal solution, we run the model multiple times with an increasing number of integer cuts [3] to produce a rank-ordered list of the optimal designs for the biosensor.

The model is applied to a case study in which we have to choose from a set of 11 receptors to recognize eight ligands in a sample. It is found that the top ten designs require only two or three receptors to optimally recognize all ligands.

[1]- J. M. Cooper and A. E. G. Cass, eds., *Biosensors: A Practical Approach.* Oxford University Press (2004).

[2]- A. T. Wright and E. V. Anslyn and J. T. McDevitt. "A Differential Array of Metalated Synthetic Receptors for the Analysis of Tripeptide Mixtures." *J. Am. Chem. Soc.* 127 (2005): 17405-17411.

[3]- C.A. Floudas. *Nonlinear and Mixed-Integer Optimization: Fundamentals and Applications.* Oxford University Press (1995).

576h Fractal Binding and Dissociation Kinetics of Prion Proteins on Biosensor Surfaces

Ajit N. M. I. Sadana[1], **Reema Taneja**[2], and Kennon C. Shelton[2]. (1) University of Mississippi, Chemical Engineering Department, University, MS 38677-1848, (2) Chemical Engineering Department, University of Mississippi, Rebel Drive, University, MS 38677-1848

A fractal analysis is presented for the binding and the dissociation of prion proteins to biosensor surfaces. Both single-and dual-fractal analysis may be used to model the binding and the dissociation kinetics. The fractal dimension provides a quantitative measure of the degree of heterogeneity on the biosensor surface. The binding and dissociation of human Clq in solution to PrP immobilized on a sensor chip surface is analyzed [11]. Quantitative relations are presented for (a) the binding rate coefficient, k as a function of the Clq in solution, (b) the dissociation rate coefficient, kd as a function of the Clq in solution, (c) the affinity, K(=k/ kd) as a function of the ratio of the fractal dimensions present in the binding and in the dissociation phase Df/Dfd, and (d) the affinity K as a function of the Clq concentration in solution. These relationships developed provide fresh physical insights into the interactions of prion proteins on biosensor surfaces and into these types of reactions in general.

627a Selectivity Control and Mechanistic Studies of Catalytic Mis Sensors

J. Will Medlin and Stephen Marshall. Department of Chemical and Biological Engineering, University of Colorado, ECCH 111, UCB 424, Boulder, CO 80501

Metal-insulator-semiconductor devices are promising sensors for low concentrations of hydrogen and other molecules. The major challenge in design and use of these sensors is in optimization of selectivity for the desired analyte. Selectivity improvement is hampered by a poor understanding of the catalytic chemistry by which these devices function. In this presentation, we will discuss our efforts to design devices for improved selectivity (by use of alloys and surface-modifying monolayer films) and to understand mechanisms for selectivity through surface characterization. Possible applications for designed sensors will be discussed.

627b Application Of VO_x/Al_2O_3 & $Fe_2(MoO_4)_3$ For Selective Catalytic Detection Of Ethanol In Multi-Component Hydrocarbon Mixtures

Joseph E. Gatt and Chelsey D. Baertsch. Chemical Engineering, Purdue University, Forney Hall of Chemical Engineering, # 2125, 480 Stadium Mall Drive, West Lafayette, IN 47907

With the recent push towards utilizing ethanol as the primary oxygenate in automotive fuel, there is a demand for sensors capable of on-site control of ethanol/gasoline blending and verification of ethanol content of fuels. Requirements for such technology generally include small size, low power consumption, and selective detection with minimal false responses from mixture components. Current sensor technologies utilize microelectromechanical systems (MEMS) to satisfy size and power constraints; however, selective detection and quantification remains an issue. A new sensing methodology is proposed based on reaction microcalorimetry in which an intrinsically selective catalyst deposited on a thin film temperature sensor is used to selectively oxidize only the target gas analyte.

For a catalytic sensor to be useful for ethanol detection in multicomponent hydrocarbon mixtures, the catalyst must show high product selectivity towards one exothermic product such as acetaldehyde, be inactive towards background gases at the sensor operating temperature, and exhibit either high rates of ethanol consumption and/or be 1st order in terms of ethanol concentration in order to allow for quantification. Important parameters for designing such a system are operating temperature and catalytic material. Catalysts which exhibit the desired characteristics at low temperatures are desirable in order to limit power consumption of a prospective device and to avoid gas phase oxidation of the more volatile components in gasoline.

Two promising catalysts for satisfying sensing requirements are VO_x/Al_2O_3 and $Fe_2(MoO_4)_3$. Acetaldehyde selectivity over VO_x/Al_2O_3 increases with increasing coverage. 84% acetaldehyde selectivity at 180 °C is achieved over a catalyst with 8 V/nm^2. Ethanol partial oxidation rates over 8 VO_x/Al_2O_3 are zero order in ethanol concentration at 180 °C, with rates equal to $8.8x10^{-9}$ mol ethanol converted/m^2 s ($1.3x10^{-6}$ mol ethanol converted/g catalyst s). In comparison, $Fe_2(MoO_4)_3$ demonstrates higher selectivity towards acetaldehyde (98%) at 180 °C with ethanol conversion rates zero order in ethanol concentration. However, the rate of ethanol consumption per gram catalyst over $Fe_2(MoO_4)_3$ is lower than 8 VO_x/Al_2O_3; equal to $8.6x10^{-7}$ mol ethanol converted/g catalyst s. Kinetic investigations involving binary mixtures of ethanol and other hydrocarbons representing key classes of compounds in gasoline have also been conducted at 180 °C. Ethanol partial oxidation rates and acetaldehyde selectivities measured during reactions with equimolar mixtures of benzene or methane with ethanol over 8 VO_x/Al_2O_3 remain unchanged. Neither benzene nor methane show any activity at 180 °C. Reactions involving equimolar mixtures of ethanol and 2-butanol over 4 VO_x/Al_2O_3 exhibit the same selectivity as seen in experiments involving each component individually, but their oxidation rates differ, with ethanol turn over rates dropping from $8.1x10^{-4}$ s^{-1} to $2.1x10^{-4}$ s^{-1}. Both VO_x/Al_2O_3 and $Fe_2(MoO_4)_3$ are promising catalysts for detection and quantification of ethanol through selective oxidation to acetaldehyde in the presence of less reactive hydrocarbons.

627d Correlation of Microhotplate Metal Oxide Sensor Response to Catalytic Fluorocarbon Decomposition Activity

Aaron Clark[1], Mauricio Pereira da Cunha[2], Bruce Segee[2], and M. Clayton Wheeler[1]. (1) Department of Chemical and Biological Engineering, University of Maine, Orono, ME 04469, (2) Department of Electrical and Computer Engineering and Laboratory for Surface Science and Technology, University of Maine, Orono, ME 04469

A hybrid microsensor system consisting of an array of microhotplate reactors/sensors and a complimentary surface acoustic wave (SAW) sensor is being developed for selective detection of fluorocarbon compounds using tetrafluoroethane (R134A) as a model compound. The microhotplate sensors simultaneously detect and decompose the analyte while the SAW depends on exposure to the hydrolysis products of the R134A for its sensing mechanism. This paper focuses on comparing the microhotplate sensing and catalytic reaction mechanisms.

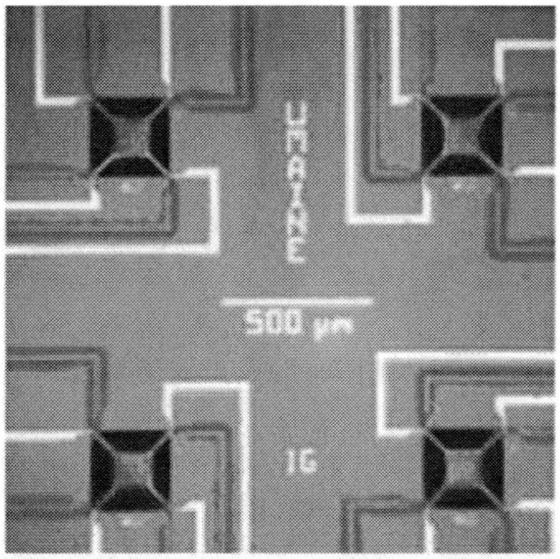

Figure 1. Four element microhotplate array

The microhotplate devices, shown in Fig. 1, were fabricated in the clean room at the University of Maine's Micro Instruments and Systems Laboratory and each consists of a suspended sandwich of multiple SiO_2/SiN layers that separate two platinum circuit element layers: 1) an embedded serpentine heater and 2) surface electrical contacts. Since microhotplates are thermally isolated from the substrate by silicon micromachining, they can be heated and cooled over a range of more than 400 °C at rates of over 10^6 °C per second. Semiconducting SnO_2 and catalytically active platinum are selectively deposited on the microhotplates using chemical vapor deposition with precursors tin tetrabutoxide and platinum acetylacetonate. Measuring the electrical response of the material over a range of operating temperatures is a way to improve the selectivity of metal oxide sensors that is commonly referred to as temperature programmed sensing (TPS).

In TPS the reaction kinetics control the surface concentrations of species such as oxygen and hydroxyl. These surface concentrations significantly affect the electrical conductivity of the sensing film. The rates depend on the sensor material as well as the temperature. The sensor sensitivity and selectivity of SnO_2 (with and without Pt catalyst doping) are correlated to the decomposition rate of R134A measured using a microreactor catalyst characterization system.

Sensitivity of a Surface Acoustic Wave Hydrogen Fluoride Sensor to Quartz Substrate Etching

*B. J. Meulendyk†, M.C. Wheeler‡, B. Segee†, and M. Pereira da Cunha†**
†Department of Electrical and Computer Engineering
‡Department of Chemical and Biological Engineering
**Laboratory for Surface Science and Technology*
University of Maine, Orono, ME, 04469, USA

1. Introduction

Hydrogen fluoride (HF) is used in many industrial processes including the manufacture of ceramics, steel, chemicals, polymers, semiconductors, and refrigerants. Acute exposure to concentrations of HF exceeding 30 ppm is immediately dangerous to human life while chronic exposure to concentrations as low as 3 ppm may significantly affect human bones and organs. In addition to process monitoring, many environmental contaminants containing fluorine could be selectively detected via HF decomposition products using hybrid sensor systems.

Surface acoustic wave (SAW) resonators have been extensively investigated for a variety of sensing applications, and are well suited for both hybrid and stand-alone sensor systems. Quartz is well suited for HF sensing with SAW resonators because the analyte reacts directly with the substrate surface, producing H_2O and SiF_4 [1, 2]. It has been suggested that SAW devices respond to HF via the desorption of these products [3, 4].

This paper shows evidence that during gas phase HF exposure to a pure SH SAW resonator, the dominant sensing mechanism reflects the detection of a condensed liquid layer on the SiO_2 surface, rather than material removal from the device's surface. At HF concentrations ranging from 1 ppm to 18 ppm, the resonator responds to the changing state of the condensed liquid layer with clear frequency shifts within minutes of exposure. Permanent frequency shifts in the SAW response are observed due to the etching of SiO_2 after the evaporation of the liquid layer when HF is removed from the test gas for longer than 3 hours. However those permanent shifts are smaller than the frequency shift response due to the presence of the condensed liquid layer.

Section 2 of this paper discusses the nature of the reaction between HF and SiO_2. In section 3 the gas delivery system is described as well as details regarding the SAW resonator, its packaging, and inclusion in the test system. Section 4 contains the experimental results and respective analysis. Section 5 concludes the paper.

2. Mechanism for vapor phase HF etching of SiO_2

The characteristics of HF etching of SiO_2 have been well documented, primarily for application in the semiconductor industry. In 1966 Holmes and Snell first reported that vapor phase and liquid phase etch rates were of comparable magnitude and attributed this effect to the formation of a liquid film on the SiO_2 surface [3], after which it is assumed that etching occurs as if the SiO_2 were in a liquid solution [4]. The reaction can be described by

$$SiO_2 + 6HF \rightarrow H_2SiF_6 + 2H_2O, \quad (1)$$

and has been represented in previous work by a diagram similar to that in Figure 1 [4]. The diagram depicts exposure of SiO_2 to HF vapor in Figure 1a; if the partial pressures of HF and H_2O are sufficiently high as described in [4], a condensed layer of HF and H_2O will form on the surface (Figure 1b). With the formation of this layer, HF and SiO_2 react, creating H_2SiF_6 and more H_2O (Figure 1c). The H_2SiF_6 will decompose via

$$H_2SiF_6 \rightarrow 2HF + SiF_4. \quad (2)$$

At a given concentration of HF, a condensed liquid layer will form and grow until the rate of transport of HF to the SiO_2 surface is balanced by the rate of desorption of SiF_4 and H_2O from the liquid layer, thereby reaching a steady state [3]. As the concentration of HF is increased in the vapor, the concentration of HF in the liquid layer will increase and the reaction rate will increase. Thus, the thickness of the liquid layer will grow until a new steady state is achieved. Likewise, decreasing the concentration of HF in the vapor will cause the layer thickness to decrease, as more H_2O will evaporate than can be generated by the reaction. Holmes and Snell determined that the vapor phase etch rates are largest when the condensed layer is thin and when temperatures are between 20°C and 35°C; the etch rate falls off quickly at temperatures greater than 40 °C [3].

In the absence of water vapor, HF has only been shown to etch SiO_2 at temperatures greater than 400°C [5]. However, Helms and Deal assert that "small amounts of water present in the ambient, on the surface, or produced as a consequence of the initial reaction… are sufficient to produce a condensed phase" [3].

The discussion above refers to HF concentrations higher than 50 ppm. However, there is spectroscopic evidence showing the formation of persistent layers of molecular water on ordered silica as well [6,7], which justifies the formation of a condensed liquid layer on the SiO_2 substrate when exposed to HF concentrations in the range of interest (1 ppm - 20 ppm).

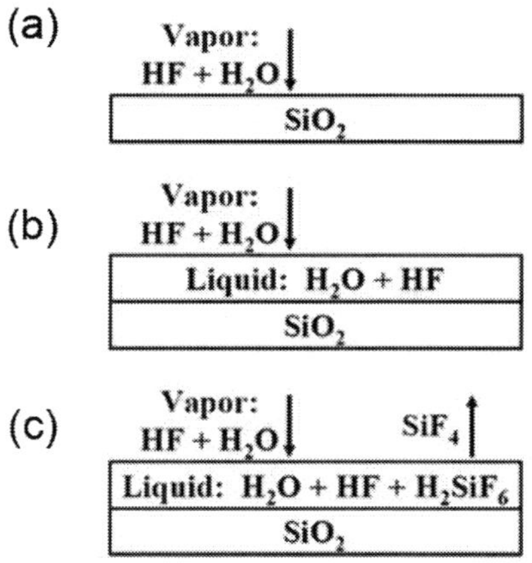

Figure 1: Diagram of the steps involved in the reaction of HF with SiO_2 [4].

3. Experimental Setup

3.1 Pure SH SAW resonator and test cell

The two-port pure SH SAW resonators used for this work were fabricated by thin film deposition with platinum electrodes on ST-cut quartz (0°, 132.75°, 90°) at the University of Maine's Laboratory for Surface Science and Technology. Figure 2 shows two pure SH SAW resonators inside a low volume test cell (≈ 1 ml) used for gas delivery experiments. The resonators are mounted on a printed circuit board (PCB) and bonded to 50Ω RF feed-throughs, which connect to the SMA pins via microstrip transmission lines on the opposite side of the PCB. Two RF switches connect the two two-port resonators to an Agilent A8753D

Network Analyzer. This arrangement allows for alternating measurement of the two resonators by a single network analyzer as they are exposed to identical test conditions.

3.2 Gas Delivery System

An automated gas delivery system, represented in Figure 3, was built to expose two-port SAW resonators to combinations of air and HF under a variety of conditions. A Parker Balston dryer and zero air generator (HPZA-3500) supplies the air. Flowing zero air through a distilled water vapor liquid equilibrium (VLE) cell allows the relative humidity (RH) to be adjusted from 0-50%. Bottled anhydrous HF, in a nitrogen carrier gas, is supplied by Matheson Tri-Gas. Mass flow controllers are used to vary gas flow rates from 40-2000 sccm and HF concentrations from 1-18 ppm. A custom, low-temperature oven was implemented using a resistive air heater operated by a proportional controller with thermistor feedback to maintain the temperature within a Styrofoam container at ±0.08 °C within the ideal range (30-40 °C) for the HF/SiO_2 reaction. A coiled length of tubing and the test cell were placed inside the oven to ensure both the test gas and SAW resonator were at the desired temperature.

4. Results & Discussion

The frequency response of the pure SH SAW resonator was measured while exposing it to different environmental conditions. Initially, experiments were performed to determine the effects of changing flow rate as well as changing humidity levels. The results indicate that changes in flow rate are insignificant, while changes in humidity are considerable.

Next, HF exposure experiments were conducted. In the first test, the resonator was exposed to HF mixed with 25% RH air in an effort to catalyze the reaction and increase the etch rate to an amount measurable by the device. The second experiment was designed to investigate the effect of increasing and decreasing HF concentrations on the condensed liquid layer. For the third experiment, the exposure time of the resonator to HF was extended from 2 hours to 6 hours to allow time for the condensed layer to approach a steady state. Also, to determine the effect of humid air, this experiment was conducted in both dry and 25% RH air. The fourth and last experiment was conducted in dry air with varying concentrations of HF, which showed that the resonator responds within a few minutes to HF concentrations as low as 1 ppm.

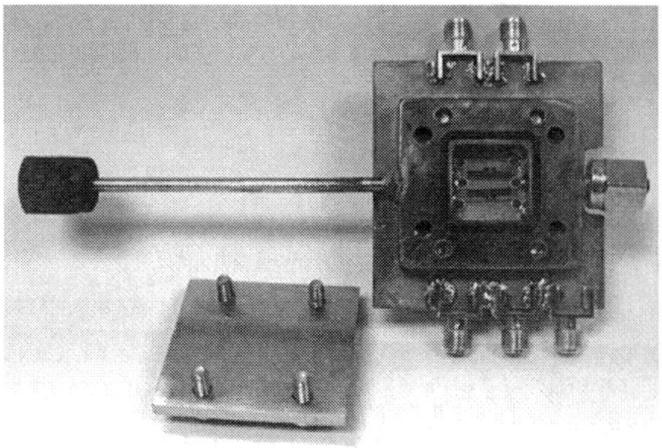

Figure 2: Two pure SH SAW resonators mounted inside the gas test cell.

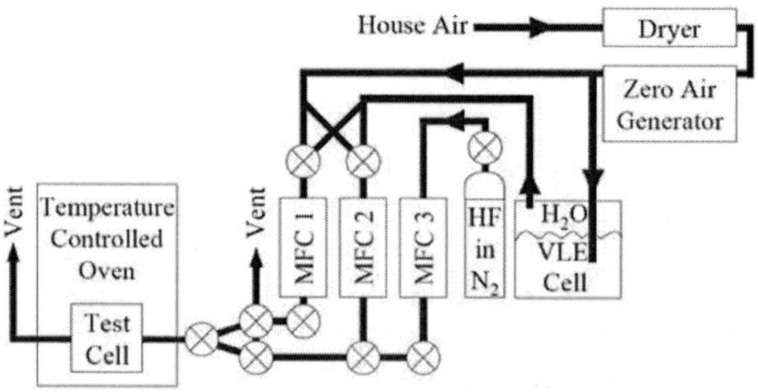

Figure 3: Representation of the gas delivery system.

4.1 *Flow Rate*

An experiment was performed in which the gas flow rates of dry and humid air were cycled to determine the effect on the SAW resonator. The results, shown in Figure 4, indicate that the effect of flow rate is insignificant, but that the influence of humidity on the resonator response is considerable. At the start of the experiment in Figure 4, a baseline frequency response at 0 sccm air flow was measured for 120 minutes. At marker 1, the device was exposed to 100 sccm of high humidity air for ≈20 minutes. At marker 2, the device was exposed to dry air for ≈100 minutes. At markers 1 & 2 the effect of moisture on the resonator is significant, causing frequency shifts of 34 kHz and 22 kHz respectively. The fact that the frequency only recovers to 65% of its original value indicates that the removal of moisture from the air does not necessarily cause the removal of strongly bonded water from the SiO_2 surface.

At markers 3, 4, 5, and 6, the flow rate of dry air was changed to 500, 1000, 1500, and 0 sccm respectively. Dry air at low flow rates continued to have little effect on the resonator response, but as the flow increased, the air did have a drying effect, exhibited by a frequency increase between markers 4 and 6.

At markers 7, 8, 9, and 10, the air was humidified to 25% RH and the flow rates were 100, 500, 1000, and 1500 sccm, respectively. After the initial wetting of the SiO_2 surface at marker 7, each subsequent change in flow rate caused a frequency shift of less than 1.5 kHz. At marker 11 the air flow was reduced to 0 sccm and the frequency recovered to its value at marker 6 and within 1 kHz of its original value.

4.2 *Humidity*

To further characterize the extent to which humidity affects the SAW resonator response, an experiment was conducted in which the relative humidity was increased and decreased between 0% and 35% RH at a constant flow rate of 100 sccm. Figure 5 shows the frequency response of the pure SH SAW device with time and changing humidity. For this experiment, the pressure in the H_2O VLE cell was set to 25 psi, creating 37.5% RH air. Dry air and humid air were mixed at eight different concentrations creating humidity levels in the range 0-35% RH. At the start of the experiment the resonator was exposed to 1500 sccm of 0% RH air to

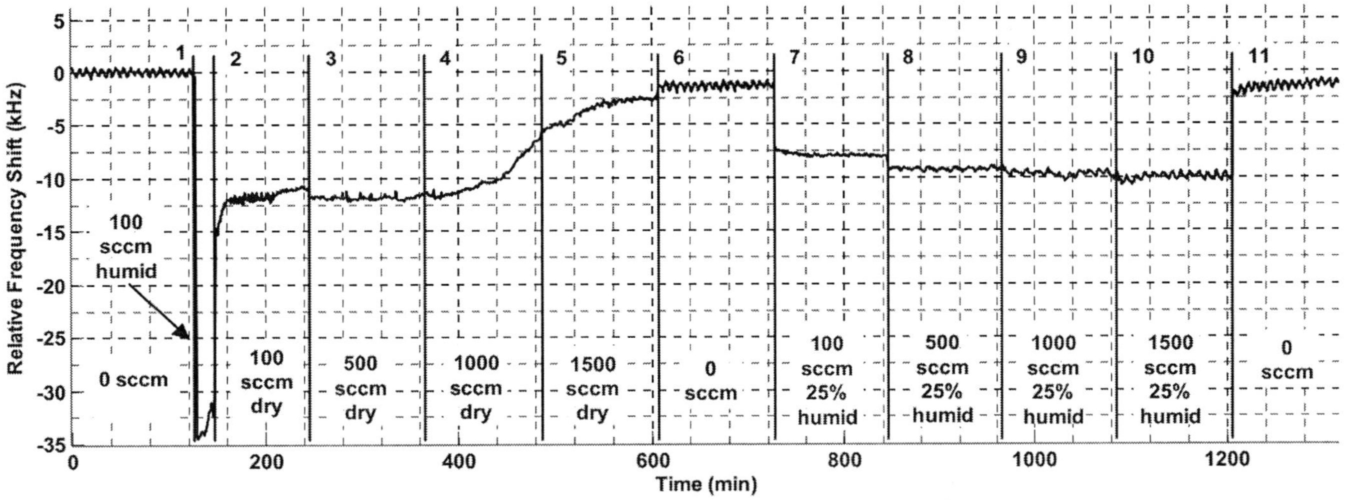

Figure 4: Response of the pure SH SAW resonator to variations in flow rate of dry and humid air at 40 °C. Device operating frequency: 293.946 MHz.

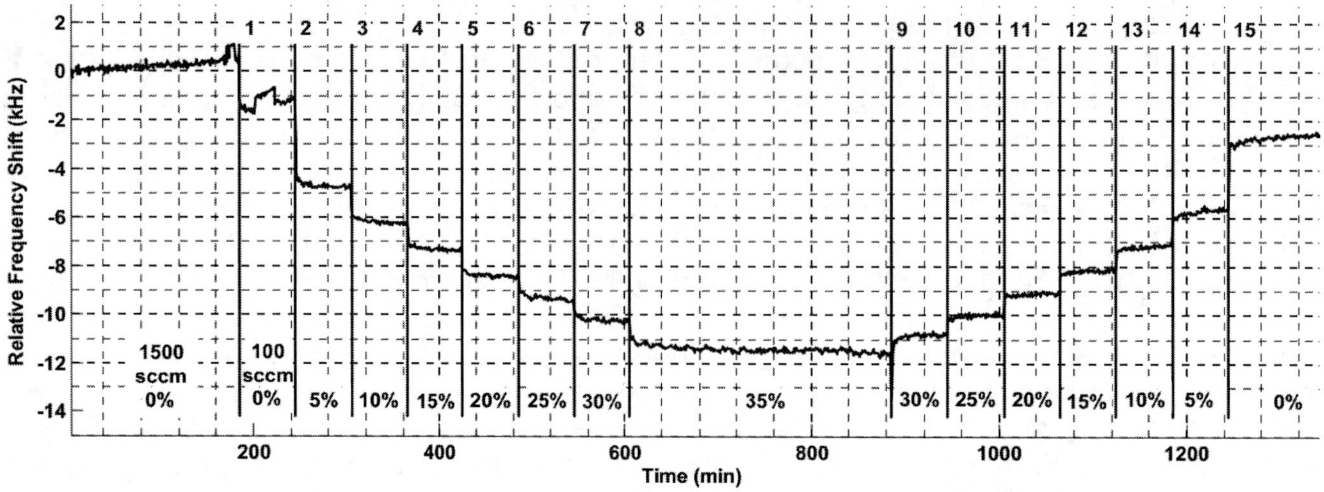

Figure 5: Response of pure SH SAW resonator to variations in humidity at 100 sccm air flow and 35 °C. Device operating frequency: 293.902 MHz.

dry the SiO_2 surface. At marker 1, the flow rate was decreased to 100 sccm to establish a baseline dry air response at this flow rate. The relative humidity of the air was increased in 5% increments every 60 minutes until 35% RH was achieved at marker 8. After 280 minutes the humidity was decreased by 5% every 60 minutes until 0% RH was achieved at marker 15. Figure 6 shows the center frequency of the resonator with respect to humidity for both wetting and drying of the SiO_2 surface. From Figures 5 and 6 it is clear that after the initial wetting of the SiO_2 surface at 5% RH, the frequency response changes linearly with changing RH, indicating that moisture content on the surface of the resonator is also changing linearly. Figure 6 also indicates that at 100 sccm, the liquid layer is not removed from the SiO_2 surface as easily as it is created.

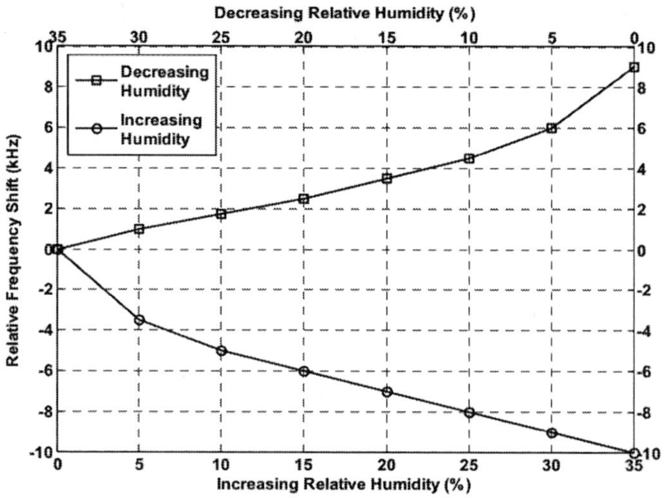

Figure 6: Center frequency of pure SH SAW resonator vs. relative humidity at 100 sccm and 35 °C. Device operating frequency: 293.902 MHz.

4.3 *HF Concentration*

The previous experiments provide insight regarding the effects of moisture on the surface of SiO_2 and its subsequent effect on the response of the SAW resonator. This knowledge is important in understanding the response of the SAW resonator when exposed to HF.

4.3.1 Varying HF concentration in 25% RH air

As described in Section 2, it is well known that water vapor is a catalyst for the reaction of HF with SiO_2. Thus, an experiment was conducted to determine the effect on the SH SAW resonator response of HF in 25% RH air. The results of this experiment are presented in Figure 7. To start, 100 sccm of 25% RH air was applied to the resonator to establish a baseline. At marker 1, anhydrous HF, mixed with humid air at a concentration of 18 ppm, was applied to the resonator. To achieve this concentration, the flow of humid air was reduced from 100 to 80 sccm and the flow of anhydrous HF and nitrogen was increased from 0 to 20 sccm, these changes correspond to a reduction in RH from 25% to 20%. From Figure 6, it is known that this reduction in RH corresponds to a frequency increase of ≈0.750 kHz. A frequency increase of this magnitude occurs just after marker 1 for ≈16 minutes and is attributed to the decrease in RH. As the reaction of HF with SiO_2 progresses, the frequency decreases at a rate of ≈45 Hz/min, indicative of a growing condensed layer.

At marker 2, the HF concentration is reduced to 0 ppm, and humid airflow is returned to 100 sccm. With this change, the frequency of the resonator drops ≈0.750 kHz (consistent with returning the humidity from 20% to 25%) before partially recovering, indicating that the sensitivity of the resonator to HF is not solely due to the permanence of etching the SiO_2 substrate but is also due to the production of the condensed layer.

At marker 3, 9 ppm HF in humid air is applied to the resonator. Again, the frequency initially increases for ≈4 minutes, but the effect of the growing condensed layer on the surface quickly dominates the resonator response, causing the frequency to decrease at a rate of ≈48Hz/min, similar to that of the 18 ppm exposure. However, after 40 minutes the rate of change reduces to ≈10 Hz/min, indicating that the rate of the surface reaction is approaching a steady state balance between water production and desorption, as described in Section 2.

At marker 4, the HF concentration is reduced to 0 ppm and, again, the response partially recovers before 1 ppm HF is applied to the resonator at marker 5. The frequency of the resonator is unchanged during this exposure, signifying that 1 ppm HF at 25% RH is enough to maintain the condensed layer that remains on the resonator surface. Humid airflow is returned to 100 sccm with 0 ppm HF at marker 6 and the frequency recovers to a value 0.750 kHz lower than the starting value, which is evidence that the resonator is capable of measuring small frequency variations due to the permanent etching of the SiO_2 substrate.

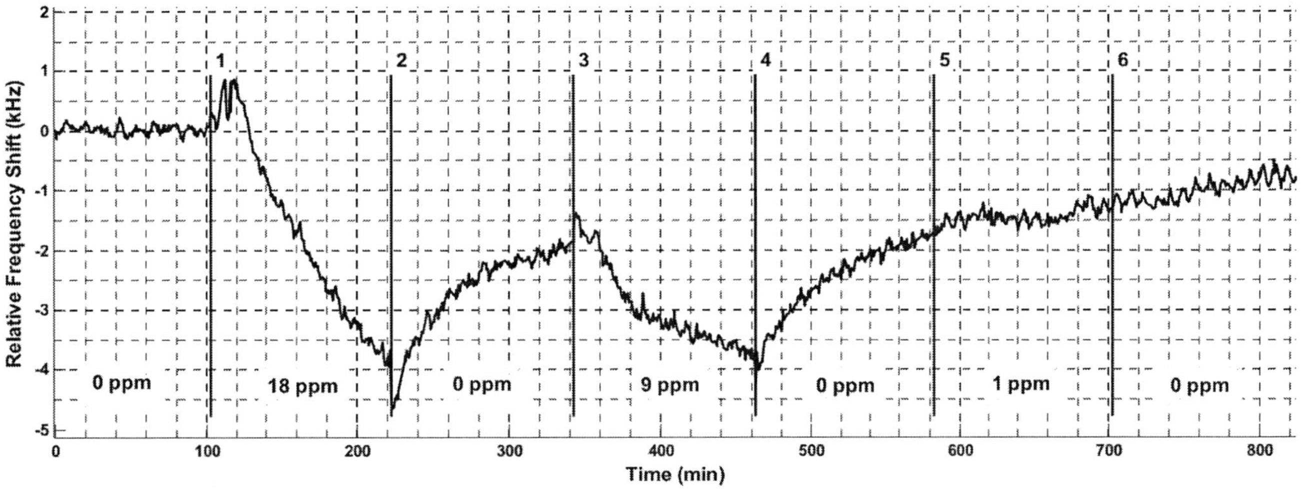

Figure 7: Response of pure SH SAW resonator to varying concentrations of HF in 25% RH air at 100 sccm and 40 °C. Device operating frequency: 293.938 MHz.

4.3.2 *The dynamic state of the condensed layer*

Figure 8 exhibits the results of an experiment conducted to verify the effects of increasing and decreasing HF concentration on the condensed liquid layer, as described in Section 2. In particular, the resonator was exposed to consecutive changes in HF concentration in 100 sccm of 25% RH air, without intermediate drying steps. To start, the response of the resonator at 0 ppm was measured. At markers 1, 2, 3, 4, and 5, HF in humid air was applied to the resonator in concentrations of 18, 12, 6, 1, and 0 ppm respectively, at two hour intervals. After the HF is applied at marker 1, the typical frequency increase is observed before the condensed layer forms, causing the frequency to decrease at a rate of ≈30 Hz/min, as it grows. At marker 2, the HF concentration is reduced to 12 ppm and the slope of the response decreases to ≈15 Hz/min, indicating that the condensed layer has slowed its growth due to the reduction of HF. At marker 3, when the concentration is reduced to 6 ppm, more water is evaporating from the surface than can be generated by the reaction of HF with SiO_2, causing the layer to diminish and the frequency to increase by 0.5 kHz. This effect is more pronounced at marker 4, when the HF concentration is further reduced to 1 ppm and the frequency increases 2 kHz in two hours. When the HF concentration is reduced to 0 ppm at marker 5, the frequency response recovers to a value ≈1.5 kHz below the starting value.

4.3.3 *Extended exposure in dry and humid air*

In the previous experiments, the resonator responses to higher concentrations of HF appeared largely linear while the responses to lower concentrations appeared logarithmic. If the resonator is indeed mostly responding to a condensed liquid layer formed by the reaction of HF with SiO_2, the resonator frequency variation would be expected to reduce to zero as the liquid layer reached its maximum thickness. Since the time required to achieve a steady state liquid layer is expected to increase with HF concentration, an experiment was conducted in which the exposure time of the resonator to 18 ppm was increased to 360 minutes in both dry and 25% RH air. The results of this experiment are presented in Figures 9a (25% RH air) and 9b (0% RH air). In both experiments the slope of the resonator response approaches zero, as expected. In humid air, the frequency increased by 0.5 kHz in 10 minutes, decreased by 1.1 kHz in the first 60 minutes and a total of 4 kHz in 360 minutes. The effect of 18 ppm HF is more pronounced in dry air, as the frequency quickly decreased 1.8 kHz in 10 minutes,

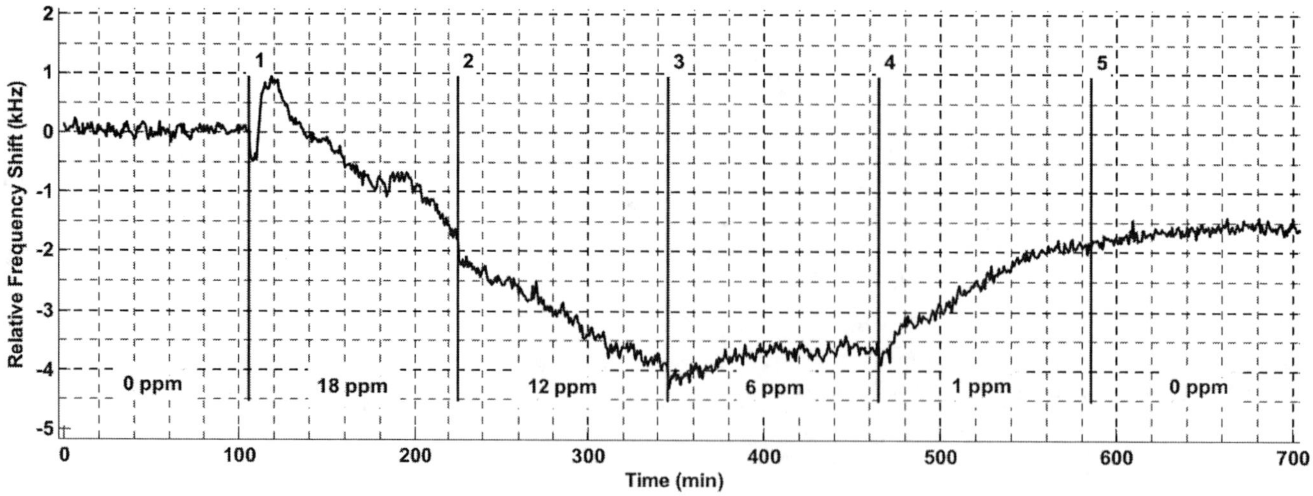

Figure 8: Response of pure SH SAW resonator to varying concentrations of HF in 25% RH air at 100 sccm and 40 °C. Device operating frequency: 293.938 MHz.

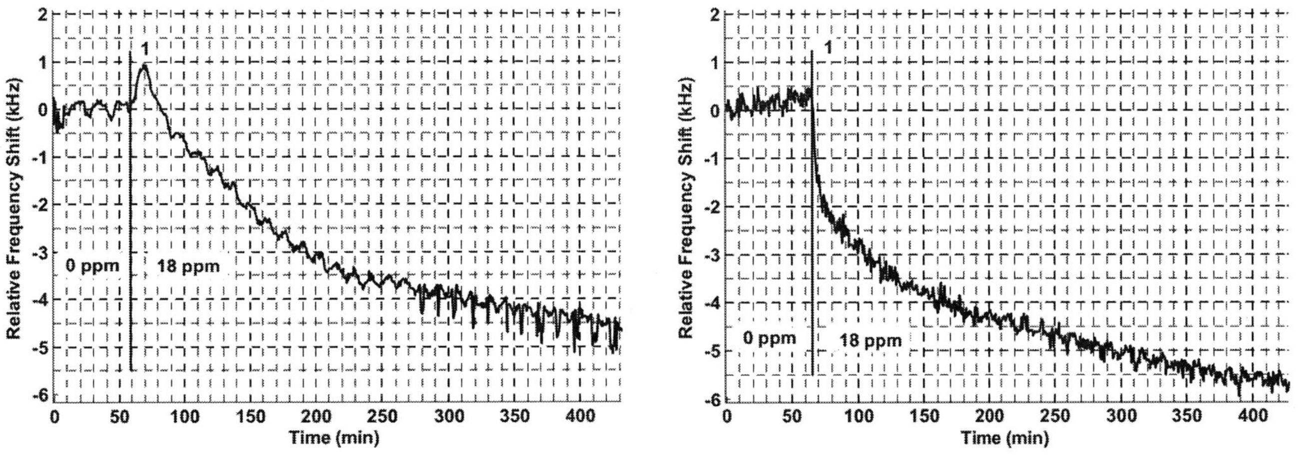

(a) (b)

Figure 9: Response of pure SH SAW resonator to 18 ppm HF in (a) 25% RH air and (b) 0% RH air at 100 sccm and 35 °C. Device operating frequency: (a) 293.896 MHz (b) 293.895 MHz.

3.1 kHz in 60 minutes, and a total of 4.8 kHz in 360 minutes. These results not only indicate that the HF/SiO_2 reaction is able to quickly produce enough water to maintain the reaction in dry air but also that the resonator is more sensitive to HF in dry air.

The difference in sensitivity can be attributed to the condensed liquid layer. In dry air, when HF is introduced it immediately reacts with the SiO_2 creating enough H_2O to be sensed by the resonator within minutes. In humid air, a thin condensed layer of H_2O already exists on the substrate surface and, when introduced, the HF must be transported though the condensed layer before reacting with the SiO_2 to create H_2SiF_6 and more H_2O. In addition to this reaction delay, the condensed layer will also grow at a lesser rate as the H_2SiF_6 increases towards equilibrium within the layer. The change in layer thickness between the final thickness due to HF and the humid baseline will be less than for a dry air baseline for which no initial condensed layer exists.

4.3.4 Varying concentration in dry air

The results of the previous experiment suggested that too much water vapor had a detrimental effect on the sensitivity of the resonator to HF. Thus, an experiment was conducted to determine the sensitivity of the resonator to HF in dry air. Figure 10 shows the response of the SAW resonator to 1 ppm HF in dry air and Figure 11 shows the response to 6, 12, and 18 ppm, with intermediate exposures to 0 ppm. To start, 100 sccm dry air at 35 °C was applied to the resonator before marker 1, in Figure 10, to obtain a baseline response. At marker 1, the resonator was exposed to 1 ppm HF and responded at an initial rate of ≈40 Hz/min before the condensed layer approached a steady state condition, causing a total frequency shift of

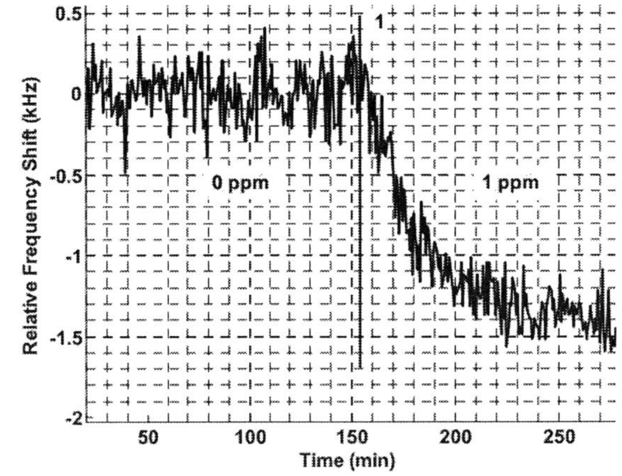

Figure 10: Response of pure SH SAW resonator to 1 ppm HF in 0% RH air at 100 sccm and 35 °C. Device operating frequency: 293.896 MHz.

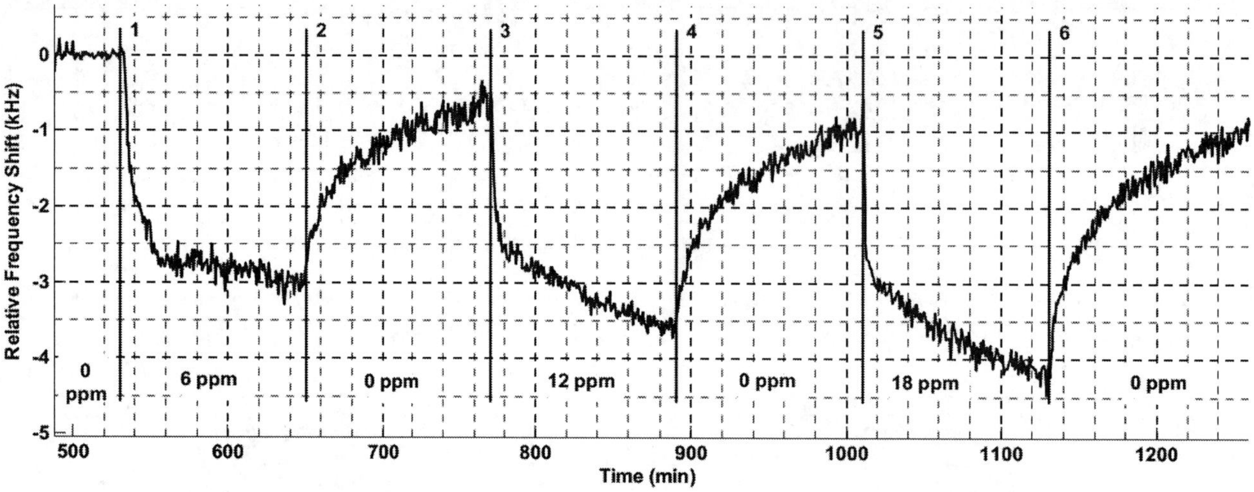

Figure 11: Response of pure SH SAW resonator to varying concentrations of HF in 0% RH air at 100 sccm and 35 °C. Device operating frequency: 293.897 MHz.

≈1.5 kHz. In Figure 11, the measurements also began with 100 sccm dry air at 35 °C. At markers 1, 2, 3, 4, 5, and 6, HF in dry air was applied in concentrations of 6, 0, 12, 0, 18, and 0 ppm respectively, for 120 minute intervals. The resonator responds to the application of 6 ppm HF with a 1.1 kHz shift in 10 minutes and a shift of 3 kHz in 120 minutes. For 12 ppm HF, the resonator response shifted by 2.5 kHz in 10 minutes and by 3.1 kHz in 120 minutes. Lastly, for 18 ppm HF, the resonator responded with a 2.5 kHz shift in 10 minutes and a shift of 3.7 kHz in 120 minutes. From Figure 11, it is evident that the response time to HF in dry air is much shorter than the recovery time.

5. Conclusions

Results reported in this paper verify that a pure SH SAW resonator fabricated on ST quartz (0° 132.75° 90°) is sensitive to a condensed liquid layer formed by the reaction of HF with SiO_2. The resonator is capable of sensing concentrations of HF as low as 1 ppm in dry air. The resonator is less sensitive to HF in 25% RH air and the lowest detectable concentration of HF is 6 ppm in the humid environment. The resonator's responses to independent changes in flow rate and humidity level were presented.

6. References

[1] P.J. Holmes, J.E. Snell, Microelectronics and Reliability, 5 (1966) 337-341.

[2] C.R. Helms, B.E. Deal, Journal of Vacuum Science and Technology A, 10 (1992) 806-811.

[3] S. Ballandras, M.H. Nadal, W. Daniau, V. Blondeau-Patissier, G. Martin, V. Eyraud, D. Hauden, B. Guicharda, L. Hairault, B. Lebret, P. Blind, Proceedings of the IEEE IFCS, (2003).

[4] V. Blondeau-Patissier, S. Ballandras, G. Lengaigne, W. Daniau, G. Martin, P. Blind, D. Hauden, M.H. Nadal, Sensors and Actuators B, 111-112 (2005) 219-224.

[5] J.M. Ballato, S.M. Laffey, R.E. Riman, J.R. Vig, IEEE IFCS, (1996).

[6] J.T. Yates Jr., Surface Science Perspectives, 565 (2004) 103-106.

[7] S. Wendt, M. Freichs, T. Wei, M.S. Chen, V. Kempter, D.W. Goodman, Surface Science, 565 (2004) 107-120.

RESPONSES OF SURFACE ACOUSTIC WAVE (SAW) SENSOR IN GAS PHASE

Zhixiong Cha
Department of Civil and Environmental Engineering
University of Utah, Salt Lake City, Utah, 84102

For polymer coated surface acoustic wave (SAW) sensors, the frequency shift due to organic vapor absorption in gas phase is much higher than the frequency shift calculated by the mass loading effect of organic vapor on polymer film. A semi-empirical equation based on fractional free volume contribution of gas vapor induced swelling effect has been introduced to describe the amplification effect of organic vapor on viscoelastic polymer films. This equation is based on an assumption that the frequency shift caused by film's viscoelastic changes due to vapor absorption is additive and proportional to frequency shift caused by mass loading effect due to vapor absorption. The frequency shift due to mass loading effect of vapor doesn't means that the initial frequency shift of the polymer film is also only due to mass loading effect of the film. But in this semi-empirical equation, the initial frequency shift by viscoelastic effect of the polymer film before vapor absorption hasn't been taken into account in the amplification factor. This discrepancy causes deviation of prediction. The equation based on perturbation theory is derived to calculate the mass loading effect of gas vapor on SAW sensor. Two equations are compared and the amplification factor in the semi-empirical equation is corrected to include the viscoelastic effect. Responses of SAW sensor to different organic vapors are calculated by the equations derived from perturbation theory and the corrected semi-empirical equation and compared with experimental data to evaluate reliability of the equations.

PURTERBATION THEORY, SEMI-EMPIRICAL EQUATION AND CORRECTIFIED SEMI-EMPIRICAL EQUATION

When surface acoustic wave (SAW) propagating on the surface of piezoelectric substrate, the mechanical impedance caused by a film coated on this substrate is the ratio of the shear stress to the particle velocity at the substrate/film interface. The surface wave complex propagation factor γ is represented by the wavenumber k_0 and attenuation α [*1*]:

$$\gamma = k_0 + j\alpha = \frac{\omega}{v_0} + j\alpha \qquad (1)$$

Where ω is the angular frequency, v_0 is the propagation velocity in uncoated film, $j=(-1)^{1/2}$.

When the frequency is held constant, external influences on such a wave (such as coating a polymer film) will raise the perturbation of the propagation factor. The normalized perturbation is [*1*]:

$$\frac{\Delta\gamma}{k_0} = \frac{\Delta\alpha}{k_0} - j\frac{\Delta v}{v_0} = \sum_{i=1}^{3} c_i Z_i \qquad (2)$$

Where Z_i is the surface mechanical impedances, c_i are normalized surface particle velocities, v_0 is the wave velocity and Δv is wave velocity change in substrate. The key representational difference between TSM devices and SAW devices is that impedance of TSM device is only one-dimensional.

The perturbation method can be used to derive the acoustic load impedance of the single film on a substrate. The SAW velocity and attenuation are related with the film's dynamic behavior (applicable for acoustically thin and thick films) [*2,3,4,5*]:

$$\frac{\Delta\gamma}{k_0}=\frac{\Delta\alpha}{k_0}-j\frac{\Delta v}{v_0}=\sum_{i=1}^{3}\frac{c_i\beta_i M_i}{\omega}\tanh(j\beta_i h) \qquad (3)$$

Where the SAW-film coupling parameter $c_i = v_{i0}^2/(4k'_0 P)$, v_{i0} is the surface particle velocity in each direction, and P is the SAW power density. In this equation k'_0 is propagation constant, $\beta_i = \omega\left(\frac{\rho - E^{(i)}/v_o^2}{M_i}\right)^{1/2}$, with M_1=M_3=G (complex shear modulus) and M_2=K (bulk modulus). β_i is the propagation constant for wave polarization direction x_i propagation across the film, with modulus $E^{(i)}$ = the ratio of stress to each different strain. Here ρ is the film density.

Normally the polymer film thickness of a SAW sensor is between 10-100 nm and is chosen depending upon the film's modulus and operating frequency [*6,7*]. Martin's defined a parameter R. When R<<1, film is acoustic thin and when R>=1, this film is acoustic thick. Here R is expressed as [*8,9*]:

$$R \approx hk_0\frac{v_0^2}{v_f^2} \qquad (4)$$

Where v_0 and k_0 are the SAW velocity and the wave vector in the substrate, respectively, h is the film thickness, and v_f is the shear velocity in the film. Wohltjen and Snow have ignored film properties and suggested a guideline that the coating thickness should be less than 1% of the acoustic wavelength [*10,11*]. Ricco give a direct guideline for the layer thickness $h \ll \lambda_0(v_f/v_0)$, here λ_0 is the acoustic wavelength [*12*].

For acoustically thin films $|\beta_i h| \ll 1$ for i =1,2,3 so that $\tanh(j\beta_i h) \cong j\beta_i h$, resulting in [*2,3*]:

$$\frac{\beta_i M_i}{\omega}\tanh(j\beta_i h) \rightarrow j\omega h(\rho - \frac{E^{(i)}}{v_o^2}) \qquad (5)$$

Thus from perturbation analysis for thin films in which surface normal gradient vanishes equation 5 reduces to [*1,2,3, 9*]:

$$\frac{\Delta\gamma}{k_o}=\frac{\Delta\alpha}{k_o}-j\frac{\Delta v}{v_o}=j\omega h\sum_{i=1}^{3}c_i(\rho-\frac{E^{(i)}}{v_o^2}) \qquad (6)$$

Where h is film's thickness; ρ is film's density. $\omega=2\pi f_0$, and f_0 is the operating frequency, which is determined by sensor's physical properties. The relative changes in the phase velocity, phase and frequency can be related by the following expression [*2,3*]:

$$\frac{\Delta v}{v_0}=\frac{\Delta\varphi}{\varphi_0}=\frac{\Delta f}{f_0} \qquad (7)$$

Where Δφ is phase shift; Δf is frequency shift to resonant frequency f_0. So the relative frequency shift caused by the film is the imaginary of equation 6.

From perturbation analysis formula, the frequency shift for a thin, nonpiezoelectric, loss less, insulating isotropic film is $\Delta\alpha/k_o$=0. Substituting the $E^{(i)}$ in terms of the Lamé constants gets classic Tiersten(Wohljten) formula [*2,13*]:

$$\frac{\Delta f}{f_o} = -\omega h[c_1(\rho - \frac{\mu}{v_o}) + c_2\rho + c_3(\rho - \frac{4\mu}{v_o^2}\frac{\lambda + \mu}{\lambda + 2\mu}) \qquad (8)$$

Where λ, μ are bulk and film shear elastic modulus, respectively; k_1, k_2 are substrate material constants; $c_2=-k_1/2\pi$, $c_3=-k_2/2\pi$ and c is the SAW-film coupling parameter. Normally c_1 is a small value because of the small particle velocities in this direction. The above equation can change to (for elastic film, high modulus)[*15*]:

$$\Delta f = (k_1 + k_2) f_o^2 h\rho \ - k_2 f_o^2 h[\frac{4\mu(\lambda + \mu)}{v_o^2(\lambda + 2\mu)}] \qquad (9)$$

In the above equations, if the film's shear modulus is very small (soft film), the following validation is always true [*14*]:

$$0.5 < \frac{\lambda' + \mu'}{\lambda' + 2\mu'} < 1 \qquad (10)$$

And if the total frequency shift is only caused by the mass loading effect of the film, then equation 9 can be reduced to [*2*]:

$$\Delta f = (k_1 + k_2) f_o^2 h\rho \qquad (11a)$$

or

$$\Delta f = (k_1 + k_2) f_o^2 \frac{m}{A} \qquad (11b)$$

Where m is the total mass of the film and A is the area of the film. For this result only mass loading effect are effective.

When the total frequency shift is caused by the mass loading effect of the film, equation 9 can be reduced to:

$$\Delta f = (k_1 + k_2) f_o^2 h\rho \qquad (12a)$$

or

$$\Delta f = (k_1 + k_2) f_o^2 \frac{m}{A} \qquad (12b)$$

Where m is the total mass of the film and A is the area of the film.

If this film is a polymer film and its bulk modulus is much larger than its shear modulus ($|K| >> |G|$), the relative frequency shift and attenuation of an acoustically thin viscoelastic film are:

$$\frac{\Delta f}{f_o} \cong -\omega h((c_1 + c_2 + c_3)\rho - \frac{(c_1 + 4c_3)}{v_o^2} G') = f_0 h(k_1 + k_2)\rho - f_0 h \frac{4k_2}{v_o^2} G' \qquad (13)$$

If this film is an elastic film and absorbs some kind of gas vapor and the partition coefficient of this gas vapor in the film is K, then the frequency shift caused by the absorption is only a mass loading effect.

$$\Delta f_v = (k_1 + k_2) f_o^2 \frac{m_v}{A} \qquad (14)$$

Where m_v is the mass of gas vapor in the film. If the film thickness change after sorption is very small after sorption, dividing equation 14 by equation 12b:

$$\frac{\Delta f_v}{\Delta f} = \frac{m_v}{m} = \frac{KC_v V}{m} = \frac{KC_v}{\rho} \quad (15a)$$

$$\Delta f_v = \frac{\Delta f K C_v}{\rho} \quad (15b)$$

Where V is the volume of film.

When the film is an acoustically thin polymer film and viscoelastic effect is taken into account, the frequency shift caused by sorption should be:

$$\Delta f_v = (k_1+k_2) f_o^2 \frac{(m_v+m)}{A} - f_o^2 h \frac{4k_2}{v_o^2} G_v' - (k_1+k_2) f_o^2 \frac{m}{A} + f_o^2 h \frac{4k_2}{v_o^2} G'$$
$$= f_o^2 h (k_1+k_2) K C_v - f_o^2 h \frac{4k_2}{v_o^2} G' a C_v \quad (16)$$

Where V is the volume of polymer film and $\Delta G'$ is the modulus change after absorption. In equation 16, the first term is only related with mass and the second term is only related with modulus, so the first term can be treated as mass loading effect of solute in polymer film and the second term can be treated as viscoelastic contribution of the polymer film affected by solute.

Grate and Zellers have derived equations for viscoelastic film based on fractional free volume contribution of gas vapor induced swelling effect [15]:

$$\Delta f_v = \Delta f C_v K \frac{1}{\rho} + \Delta f C_v K (\frac{f_l}{\rho_l})(\frac{A_{SAW}}{\alpha_{th}}) \quad (17)$$

or

$$\Delta f_v = \left(1 + \frac{f_l \rho}{\rho_l} \frac{A_{SAW}}{\alpha_{th}}\right)(\Delta f C_v K \frac{1}{\rho}) = n \Delta f C_v K \frac{1}{\rho} \quad (18)$$

Where n= $[1+(f_l \rho / \rho_l)(A_{SAW}/\alpha_{th})]$ and n is the amplification coefficient, f_l and ρ_l are vapor-dependent free volume fraction of liquefied gas vapor and liquefied gas vapor density, respectively. A_{SAW} is the kilohertz change in frequency per degree change in temperature per kilohertz coating on the device, and α_{th} is the thermal expansion coefficient of the polymer.

This equation is based on an assumption that the frequency shift caused by film's viscoelastic changes due to vapor absorption is additive and proportional to frequency shift caused by mass increase due to vapor absorption. This means the viscoelastic contribution should be taken into account when calculating the initial frequency shift Δf before sorption. For this case, one should use equation 13 to calculate Δf instate of equation 14. But the derivation of equation 18 based on equation 14, which only includes mass loading effect. So although equation 18 has the similar terms to equation 16, the derivation of this equation is not consistent when its assumption is not correct.

When Grate and Zellers [*15*] are using equation 18, although their derivation process is inconsistent, their assumption of mass loading plus viscoelastic effects is right. If the first term of their equation is only related with mass changes, their equation is still right. The mass loading effect portion of Δf is:

$$\Delta f_m = f_0^2 h (k_1+k_2)\rho = \Delta f \frac{(k_1+k_2)\rho}{(k_1+k_2)\rho - \frac{4k_2}{v_o^2} G'} = \Delta f \frac{1}{1 - \frac{4k_2}{v_o^2} \frac{G'}{(k_1+k_2)\rho}} \quad (19)$$

Substituting equation 19 into equation 18, the frequency shift caused by solute absorption is:

$$\Delta f_v = \Delta f \frac{1}{1-\frac{4k_2}{v_o^2}\frac{G'}{(k_1+k_2)\rho}} C_v K \frac{1}{\rho} + \Delta f C_v K \frac{f_l}{\rho_l}\frac{A_{SAW}}{\alpha_{th}} = \left(\frac{\rho}{\rho'} + \frac{f_l \rho}{\rho_l}\frac{A_{SAW}}{\alpha_{th}}\right)\Delta f C_v K \frac{1}{\rho} \quad (20)$$

The amplification factor n is:

$$n = \frac{\rho}{\rho'} + \frac{f_l \rho}{\rho_l}\frac{A_{SAW}}{\alpha_{th}} > 1 + \frac{f_l \rho}{\rho_l}\frac{A_{SAW}}{\alpha_{th}} \quad (21)$$

because $\rho' = \rho - \frac{4k_2 G'}{(k_1+k_2)v_o^2} < \rho$.

Comparing equation 21 with equation 18, the only difference is that the density ρ is changed to ρ′. If the assumption of SAW sensor's response is mainly due to mass loading and viscoelastic effects, and if their effects are additive, the second term of equation 18 is correct because it is derived from experimental data. At high frequency, most rubbery polymers' shear modulus is in the range of MPa to GPa, but the exact values are not available for most polymers. For a polymer film with shear modulus real value of 1 GPa and density of $1000 kg/m^3$ ($1 g/cm^3$), if this polymer film is coated on YX-cut quartz, the new $\rho' = \rho - \frac{4k_2 G'}{(k_1+k_2)v_o^2}$ =722 kg/m^3. The amplification factor n is determined by properties of polymer film and substrate. Grate et al. [*15*] have tested the A_{saw} values of polymer films on ST-cut quartz, but they have not tested A_{saw} values for polymer films on other substrates. The amplification factor n can be derived if G′ value is available. From equation 13, the amplification factor n equals:

$$\Delta f_v = f_o^2 h(k_1+k_2)KC_v - f_o^2 h \frac{4k_2}{v_o^2} G' bKC_v$$

$$= \frac{\Delta f K C_v}{\rho}\frac{(k_1+k_2)\rho}{(k_1+k_2)\rho - \frac{4k_2}{v_o^2}G'} - \frac{\Delta f K C_v}{\rho}\frac{\frac{4k_2}{v_o^2}G'}{(k_1+k_2)\rho - \frac{4k_2}{v_o^2}G'} b\rho = n\frac{\Delta f K C_v}{\rho} \quad (22)$$

Where

$$n = \frac{\rho}{\rho - \frac{4k_2 G'}{(k_1+k_2)v_o^2}} + \frac{\rho b}{1 - \frac{(k_1+k_2)\rho v_o^2}{4k_2 G'}} \ . \quad (23)$$

Comparing equation 23 and equation 18, the A_{saw} is:

$$\frac{f_l \rho}{\rho_l}\frac{A_{SAW}}{\alpha_{th}} = \frac{\rho b}{1 - \frac{(k_1+k_2)\rho v_o^2}{4k_2 G'}} \quad (24)$$

The modulus contribution is determined by b (related with solute-polymer interactive properties) and shear modulus, also determined by substrate constants.

RESPONSES OF SAW SENSOR IN GAS PHASE

Partition coefficient K and LSER Parameters

The partition coefficient K is defined as the ratio of the analyte concentration in the gas to the resulting concentration of analyte in the polymer solid. It is usually modeled as a constant, the Henry's law constant. The LSER method describes k as a linear combination of five mechanistic interactions between a sorbed (adsorbed or absorbed vapor phase) analyte and a sorbent solvent, both at equilibrium, as described below [16,17,18,19,20].

$$\log K = c + rR_2 + s\pi_2^H + a\sum\alpha_2^H + b\sum\beta_2^H + l\log L^{16} \qquad (25)$$

Where $K= C_s/C_v$ and R_2, π_2^H, $\sum\alpha_2^H$, $\sum\beta_2^H$ and $\log L^{16}$ are called solvation parameters for the solute gas. The coefficients describing these interactions may be determined experimentally by a number of methods such as gas-liquid chromatography (GLC), inverse gas chromatography (IGC) or other methods. Grate and Zellers [15,67] have developed a method of determining LSER parameters from SAW vapor sensor response data that directly takes account of viscoelastic changes in the polymeric coating films.

If the sorption behavior follows a dual-mode sorption model for hydrogen bond acid polymers interacting with basic gas vapors, then the sorption equation can be described by a bimode sorption equation of the same form used for glassy state polymers [21]. For the case of SAW nonlinear sorption, the weight percentage of solute in the polymer is also more than 10% due to specific interaction although the gas-phase concentration is low.

$$C_f = KC_v + C''\frac{BC_v}{1+BC_v} \qquad (26)$$

$K=C_f/C_v$,

Where K is the partition coefficient, C″ and B come from C′ and b. Using the recent experimental data [13,18], if the amplification factors are constants, the values for K, B, and C" of several polymer gas pairs are regressed.

Calculation of Responses

The C++ simulation program was composed to estimate responses of SAW sensor to individual gas or simple gas mixtures of less than six components.The calculation of sensor's responses to gas mixture has employed the impedance method with the semiempirical method, and the simulation program based on the following assumptions: a. The frequency shift of SAW sensor to analyte sorption only includes mass loading effect and viscoelastic effect. b. The shear modulus values of polymer film and thermal expansion coefficient values of free volume of polymer are estimated when these values are not available. The interactive energies of sorption are also estimated if they are not available. Semiempirical equations are used to correct estimated values. c. Although ultra-thin film effect can influence shear modulus and partition coefficient values, the influence is very difficult to determine. The input and output interfaces of simulation program are shown in Figure 1 and 2.

Tables 1- 4 compare the literature data with responses calculated by theoretical equations. Also the semi-empirical equation has been used to correct the results. Four typical polymer films are selected, FPOL, PIB, PECH and OV25. Responses of SAW sensors with these films in gas vapors listed in Tables 1- 4 are almost linear. The experimental data of different research groups are in the same magnitude, although their experimental conditions are not the same, so the experimental sensitivities in Figure 3-6 (corresponding to Table 1- 4) are numerical average values. In the simulation program, shear modulus and changes of modulus to analyte sorption and temperature deviation are estimated. There will be 5% errors when shear modulus changes.

Strong hydrogen bonding acidic polymers with low glass transition temperatures and shear moduli are not readily available. The nonlinearity of the response curve can be caused by nonlinear modulus change or by sorption behavior that does not obey Henry's law due to high analyte concentration in polymer film. For basic vapors leading nonlinear calibration curves on strong hydrogen bond acid polymers, the sensitivity was taken from the response at the lowest vapor concentration where the LSER method is still valid. The nonlinear responses of strong hydrogen bond acid polymer coated SAW sensor exposed in basic gas vapor are calculated and compared with literature data. The results are listed in Table 5. The calibration curve of the SAW sensor with hydrogen bonding acid polymer film to basic gas vapor is very similar to calibration curve of SAW sensor with molecular recognition sites. The adsorption of analyte in cavitands always associated with bulk absorption derived by the general dispersion interaction [22].

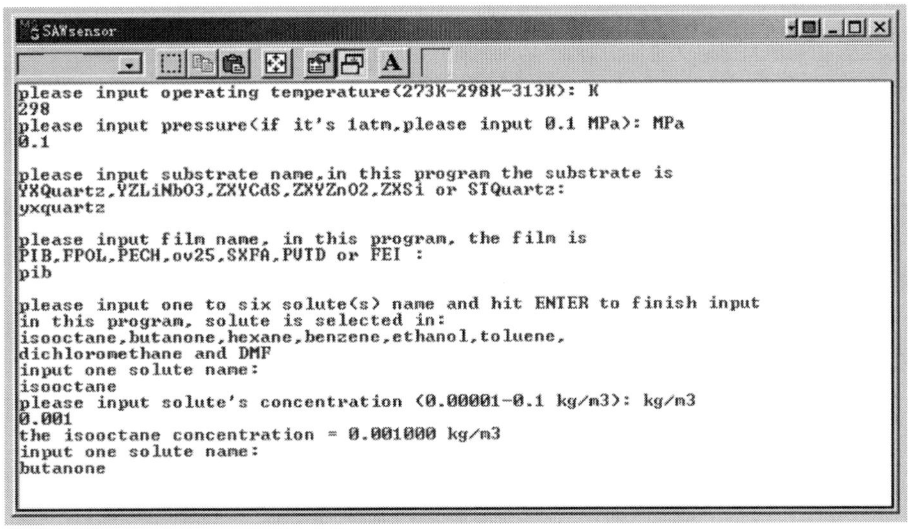

1. Input interface of simulation program

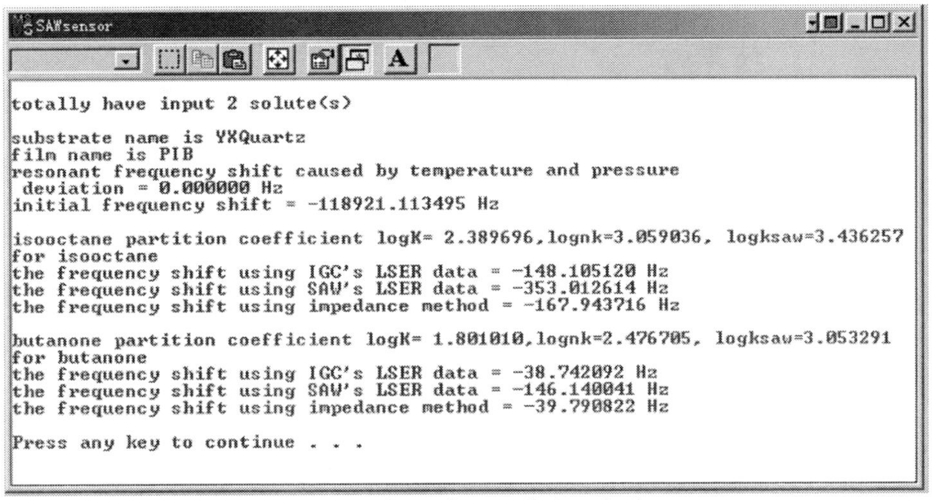

Figure 2. Output interface of simulation program

Table 1. Sensitivities of FPOL coated SAW sensors.

Table 1. Sensitivities of FPOL coated SAW sensors.

SAW	Operating frequency (MHz) ,Temp(K)	Film thickness KHz	Sensitivity Hz/(μg/l)	Normalized to 200MHz and 200K film thickness Hz/(μg/l)	Simulation Results Hz/(μg/l)
1	200 MHz 298K	250 KHz	toluene 0.28 2-butanone 0.711 1-butanol 2.91 1,2-dichloroethane 0.073	toluene 0.224 2-butanone 0.568 1-butanol 2.33 1,2-dichloroethane 0.058	toluene 0.20 isooctane 0.023 benzene 0.02 2-butanone0.86
2	200 MHz 298K	100 KHz	toluene 0.137 isooctane 0.0148 2-butanone 0.66	toluene 0.274 isooctane 0.0286 2-butanone 1.32	
3	158 MHz 308 K	222 KHz	toluene 0.123 Isooctane 0.0114 2-butanone 0.383 1-butanol 1.06 1,2-dichloroethane 0.041	toluene 0.111 isooctane 0.0103 2-butanone 0.345 1-butanol 0.955 1,2-dichloroethane 0.041	
4	250 MHz 298K	195 KHz	toluene 0.122 Isooctane 0.012 2-butanone 0.338 benzene 0.038	toluene 0.125 isooctane 0.012 2-butanone 0.34 benzene 0.039	

1.ST quartz with aluminum metallization and thin silicon dioxide[24]; 2.ST quartz with aluminum metallization and thin silicon dioxide[18]; 3. ST quartz coated with silicon dioxide [13]; 4. ST quartz [20,23].

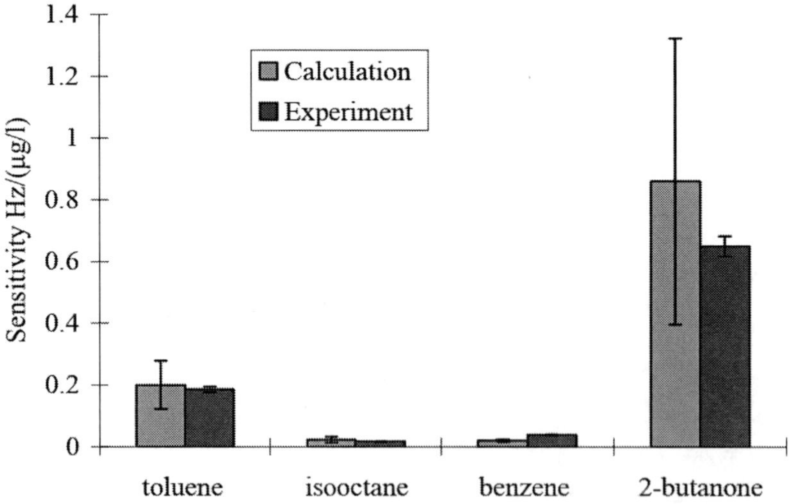

Figure 3. Sensitivities of FPOL coated SAW sensors (The experiment sensitivities are average values of data from 13,18,20,23,24)

Table 2. Sensitivities of PIB coated SAW sensors.

SAW	Operating frequency (MHz),Temp(K)	Film thickness KHz	Sensitivity Hz/(μg/l)	Normalized to 200MHz and 200K film thickness Hz/(μg/l)	Simulation Results Hz/(μg/l)
1	200 MHz 298K	280 KHz	toluene 0.838 isooctane 0.40 2-butanone 0.112 1-butanol 0.573 1,2-dichloroethane 0.159	toluene 0.599 isooctane 0.286 2-butanone 0.08 1-butanol 0.409 1,2-dichloroethane 0.113	toluene 0.51 isooctane 0.247 benzene 0.097

2	200 MHz 298K	100 KHz	toluene 0.248 isooctane 0.126 2-butanone 0.042 DMF 8.92	toluene 0.496 isooctane 0.252 2-butanone 0.084	hexane 0.11 2-butanone 0.063
3	158 MHz 291 K	180-200 KHz	isooctane 0.275 2-butanone 0.083 1-butanol 0.215 benzene 0.219	isooctane 0.28 2-butanone 0.085 1-butanol 0.22 benzene 0.22	
4	158 MHz 298K	195 KHz	toluene 0.421 Isooctane 0.193 2-butanone 0.062 benzene 0.131 hexane 0.014	toluene 0.432 isooctane 0.194 2-butanone 0.063 benzene 0.132 hexane 0.014	
5	250 MHz 298 K	225 KHz	toluene 0.875 2-butanone 0.194 1-butanol 0.484 benzene 0.314	toluene 0.78 2-butanone 0.172 1-butanol 0.43 benzene 0.28	

1.ST quartzwith aluminum metallization and thin silicon dioxide [24]; 2.ST quartzwith aluminum metallization and thin silicon dioxide[18]; 3.ST quartz [20,23]; 4.ST quartz [19].

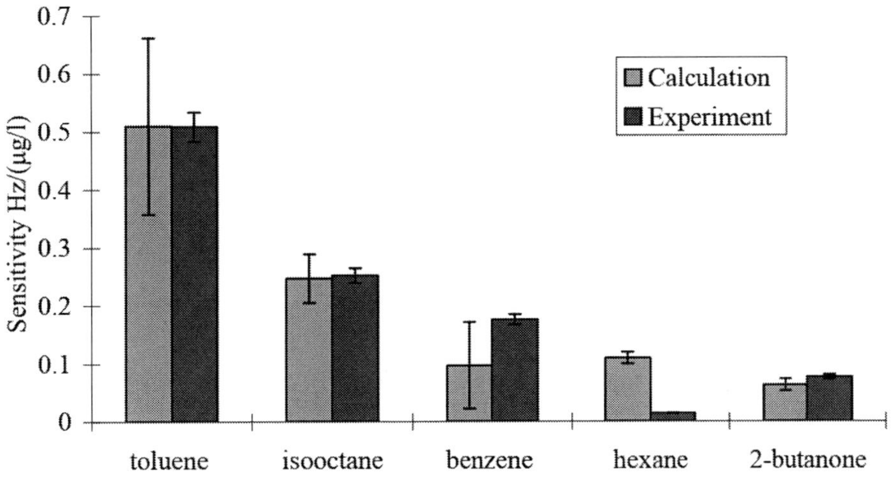

Figure 4. Sensitivities of PIB coated SAW sensors (The experiment sensitivities are average values of data from 18,19,20,23,230

Table 3. Sensitivities of PECH coated SAW sensors.

SAW	Operating frequency (MHz) Temp(K)	Film thickness KHz	Sensitivity Hz/(μg/l)	Normalized to 200MHz and 200K film thickness Hz/(μg/l)	Simulation Results Hz/(μg/l)
1	200 MHz 298K	254 KHz	toluene 0.644 isooctane 0.043 2-butanone 0.332 1-butanol 0.874 1,2-dichloroethane 0.306	toluene 0.515 isooctane 0.035 2-butanone 0.266 1-butanol 0.70 1,2-dichloroethane 0.244	toluene 0.61 isooctane 0.045 benzene 0.38 2-butanone 0.31
2	158 MHz 291 K	180-200 KHz	isooctane 0.038 2-butanone 0.245 1-butanol 0.864 benzene 0.201	isooctane 0.038 2-butanone 0.246 1-butanol 0.865 benzene 0.20	

2	158 MHz 298K	201 KHz	isooctane 0.027 2-butanone 0.159 benzene 0.147 hexane 0.012	isooctane 0.027 2-butanone 0.159 benzene 0.147 hexane 0.012	
4	250 MHz	225 KHz	toluene 0.698 2-butanone 0.598 1-butanol 1.23 benzene 0.318	toluene 0.62 2-butanone 0.53 1-butanol 1.09 benzene 0.282	

1.ST quartz with aluminum metalization and thin silicon dioxide [24]; 2.ST quartz with silicon dioxide [25]; 3.ST quartz [20,23]; 4.ST quartz [19].

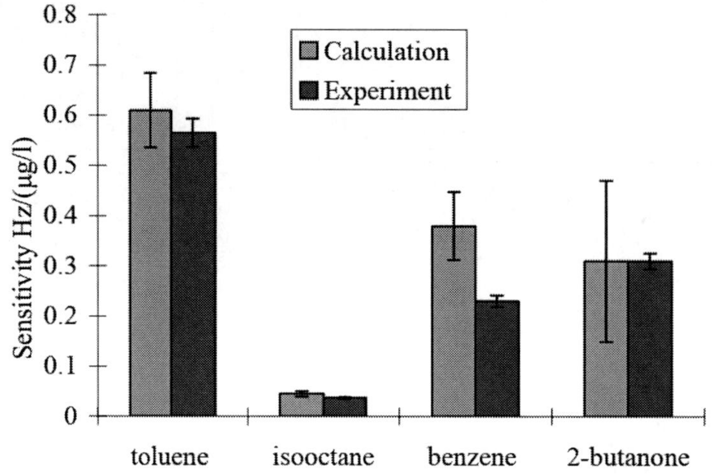

Figure 5. Sensitivities of PECH coated SAW sensors (The experiment sensitivities are average values of data from 19,20,23,24,25)

Table 4. Sensitivities of OV25 coated SAW sensors.

SAW	Operating frequency (MHz),Temp (K)	Film thickness KHz	Sensitivity Hz/(μg/l)	Normalized to 200MHz and 200K film thickness Hz/(μg/l)	Simulation Results Hz/(μg/l)
1	200 MHz 298K	100 KHz	toluene 0.468 isooctane 0.06 2-butanone 0.148	toluene 0.936 isooctane 0. 12 2-butanone 0.256	toluene 0.59 isooctane 0.069 2-butanone 0.185 hexane 0.035
2	158 MHz 298K	220 KHz	toluene 0.58 Isooctane 0.082 2-butanone 0.169 benzene 0.213 hexane 0.04	toluene 0.53 isooctane 0.076 2-butanone 0. 153 benzene 0.194 hexane 0.036	

1.ST quartz with aluminum metalization and thin silicon dioxide[24]; 2.ST quartz [20,23]

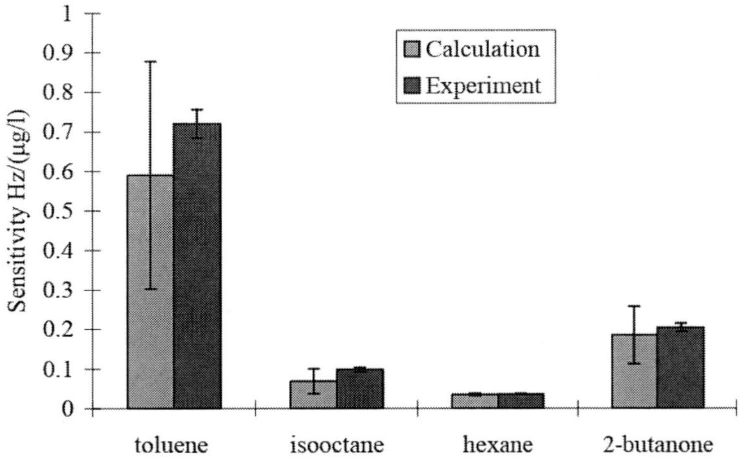

Figure 6. Sensitivities of OV25 coated SAW sensors (The experiment senstivities are average values of data from 20,23,24)

Table5. Responses of hydrogen bond acid polymer coated SAW sensor in dimethylformamide (DMF) vapor.(The experimental data are from reference 18.)

		DMF vapor concentration (μg/l)			
		0	100	400	2000
Frequency shift [a] (Hz)	Experimental	0	20000	35000	57500
	calculation	0	22200	34750	62700
Frequency shift [b] (Hz)	Experimental	0	15000	32000	65000
	calculation	0	16460	35300	68830

a.200MHz resonant frequency, film thickness is normalized to 250KHz SXFA film.
b.200MHz resonant frequency, film thickness is normalized to 250KHz FPOL film.

CONCLUSION

The results of theoretical calculations are compared with literature data. The mass-loading and viscoelastic contributions of polymer film are the most important effects responsible for frequency shift of SAW sensor. Models used or developed in this thesis can precisely predict responses of SAW sensor to single gas vapor or simple gas mixture. The dual-mode sorption model can describe absorption behavior of strong hydrogen bond polymers to basic gas vapors. Although specific interactions can enhance sensitivities of sensor to gas vapor, the response curve is normally nonlinear, and this will complicate molecular recognition and quantitative calculation.

REFERENCES

[1] Shen, C.; Shen, Y.; Wu, L. "Viscoelastic Properties of Polymer Films on Surface Acoustic Wave Organophosphorous Vapor Sensors"; *Journal of Material Science*, **2002**, 37, 295-301.

[2] Martin, S.J.; Frye, G.C.; Senturla, S.D. "Dynamics and Response of Polymer-Coated Surface Acoustic Wave Devices: Effect of Viscoelastic Properties and Film Resonance"; *Analytical Chemistry,* **1994**, 66(14), 2201-2219.

[3] DeMaggio, G. B.; Frieze, W. E.; Gidley, D. W.; Zhu, Ming; Hristov, H. A.; Yee, A. F. "Interface and Surface Effects on the Glass Transition in Thin Polystyrene Films"; *Physical Review Letters,* **1997**,78(8),1524-1527.

[4] Martin, S.J.; Frye, G.C. "Surface Acoustic Wave Response to Changes in Viscoelastic Film Properties"; *Applied Physics Letters*, **1990**,57, 1867-1869.

[5] Auld, B.A. *Acoustic Fields and Waves in Solids*;Vol 2; Wiley Interscience: New York, **1973**.
[6] Ballantine, D.S.; White, R.M.; Martin, S.J.; Ricco, A.J.; Zellers, E.T.; Frye, G.C.; Wohltjen, H. *Acoustic Wave Sensors*; Academic Press: **1997**.
[7] Martin, S.J.; Bandey, H.L.; Cernosek, R.W.;Hillman, A.R.; Brown, M.J. "Equivalent-Circuit Model for the Thickness-Shear Mode Resonator with a Viscoelastic Film Near Film Resonance"; *Analytical Chemistry,* **2000**, 72(1), 141-149.
[8] Gedde, U. W. *Polymer Physics*; Chapman and Hall: London, **1995**.
[9] Weiss, M.; Welsch, W.; Schickfus, M.V.; Hunklinger,S. "Viscoelastic Behavior of Antibody Films on a Shear Horizontal Acoustic Surface Wave Sensor"; *Analytical Chemistry*, **1998**, 70(14), 2881-2887.
[10] Bodenhöfer, K.; Hierlemann, A.; Noetzel, G.; Weimar, U.; Göpel, W. "Performances of Mass-Sensitive Devices for Gas Sensing: Thickness Shear Mode and Surface Acoustic Wave Transducers"; *Analytical Chemistry,* **1996**, 68(13), 2210-2218.
[11] Snow, A.W.; Wohltjen, H. "Poly(ethylene maleate)-cyclopentadiene: a model reactive polymer-vapor system for evaluation of a SAW microsensor"; *Analytical Chemistry,* **1984**, 56(8), 1411-1416.
[12] Ricco, A.J. *Electrochem.Soc. Interface*, **1994**,3(4), 38-44.
[13] Grate, .W.; Snow, A.; Ballantine, D.S.; Wohltjen, H.;Abraham, M.H.; Mcgill, R.A.; Sasson, P. "Determination of Partition Coefficients From Surface Acoustic Wave Vapor Sensor Responses and Correlation with Gas-Liquid Chromatographic Partition Coefficients"; *Analytical Chemistry,* **1988**,60(9),869-875.
[14] Wohltjen, H. *Sensor and Acoutors*, **1984**,5,307-325.
[15] Grate, J.W.; Zellers, .T. "The Fractional Free Volume of the Sorbed Vapor in Modeling the Viscoelastic Contribution to Polymer-Coated Surface Acoustic Wave Vapor Sensor Responses"; *Analytical Chemistry,* **2000**, 72(13), 2861-2868.
[16] Grate, J.W.; Abraham, M.H. "Solubility Interactions and the Design of Chemically Selective Sorbent Coatings for Chemical Sensors and Arrays"; *Sensors and Actuators B*, **1991**, 3, 85-111.
[17] Abraham, M.H. "Scales of Solute Hydrogen-Bonding: Ttheir Construction and Application to Physicochemical and Biochemical Processes"; *Chemical Society Reviews*, **1993**, 22, 73-84.
[18] Grate, J.W.; Patrash, S.J.; Kaganove, S.N.; Wise, B.M. "Hydrogen Bond Acidic Polymers for Surface Acoustic Wave Vapor Sensors and Arrays"; *Analytical Chemistry*, **1999**, 71(5), 1033-1042.
[19] Zellers, E.T.; Park, J.; Hsu, T.; Groves, W.A. "Establishing a Limit of Recognition for a Vapor Sensor Array"; *Analytical Chemistry,* **1998**, 70(19), 4191-4201.
[20] Park, J.; Groves, W.A.; Zellers. E.T. "Vapor Recognition with Small Arrays of Polymer-Coated Microsensors. A Comprehensive Analysis";. *Analytical Chemistry*, **1999**, 71(17), 3877-388.
[21] Lim, S.Y.; Tsotsis, T.T.; Sahimi, M. "Molecular Simulation of Diffusion and Sorption of Gases in An Amorphous Polymer"; *Journal of Chemical Physics*, **2003**,119(1),496-504.
[22] Grate, J.W.; Patrash, S.J.; Abraham, M.H.; My Du, C. "Selective Vapor Sorption by Polymers and Cavitands on Acoustic Wave Sensors: Is This Molecular Recognition"; *Analytical Chemistry,* **1996**, 68(5), 913-917.
[23] Zellers, E.T.; Batterman, S.A.; Han,M.; Patrash,S. J. "Optimal Coating Selection for the Analysis of Organic Vapor Mixtures with Polymer-Coated Surface Acoustic Wave Sensor Arrays"; *Analytical Chemistry*, **1995**,67(6) ,1092-1106.
[24] Grate, J.W.; Klusty, M.; McGill, R.A.; Abraham, M.H.; Whiting, G.; Andonian-Haftvan, J. "The predominant role of swelling-induced modulus changes of the sorbent phase in determining the responses of polymer-coated surface acoustic wave vapor sensors"; *Analytical Chemistry*, **1992**, 64(6), 610-624.
[25] Zellers, E.T.; Han, M. "Improved Method for the Determination of the Oxygen Isotopic Composition of Cellulose"; *Analytical Chemistry*, **1996**, 66(14), 2409-2418.

AUTHOR INDEX

AUTHOR INDEX